# My Appetite FOR DESTRUCTION

### SEX & DRUGS & GUNS N' ROSES

## STEVEN ADLER

### WITH
### LAWRENCE J. SPAGNOLA

!t

*itbooks*

AN IMPRINT OF HARPERCOLLINS PUBLISHERS

# *it***books**

MY APPETITE FOR DESTRUCTION. Copyright © 2010 by Steven Adler. All rights reserved. Printed in the United States of America. No part of this book may be used or reproduced in any manner whatsoever without written permission except in the case of brief quotations embodied in critical articles and reviews. For information address HarperCollins Publishers, 10 East 53rd Street, New York, NY 10022.

HarperCollins books may be purchased for educational, business, or sales promotional use. For information please write: Special Markets Department, HarperCollins Publishers, 10 East 53rd Street, New York, NY 10022.

FIRST IT BOOKS PAPERBACK PUBLISHED 2011.

*Designed by Jaime Putorti*

The Library of Congress has catalogued the hardcover edition of this book as follows:

Adler, Steven, 1965–
My appetite for destruction : sex, and drugs, and Guns n' Roses / Steven Adler with Lawrence J. Spagnola. — 1st ed.
     p.   cm.
   ISBN 978-0-06-191711-0
   1. Adler, Steven, 1965–   2. Rock musicians—United States—Biography.
I. Spagnola, Lawrence J.   II. Title.
   ML420.A275A3  2010
   786.9'166092—dc22
   [B]
                                                           2010016979

ISBN 978-0-06-191712-7 (pbk.)

11  12  13  14  15   DIX/WCF   10 9 8 7 6 5 4 3 2 1

To my grandmother "Big Lilly,"
my grandfather "Stormin' Norman,"
and to my beloved wife, Carolina, whose
love and support made this book possible.

To the millions of faithful Guns N' Roses
fans all over the world,
I thank you for your eternal devotion.

Special thanks to:
the Adler family,
the Ferreira family,
the Hudson family,
the Canter family,
Steve Sprite,
Dr. Drew Pinsky,
Dr. Charles Sophy,
Bob Forrest,
Ronald "Ronnie My Boy" Schneider,
Chris Green,
Robert Espinoza,
James Vanderweilen, and
Brad Server.

And lastly to my dogs
Shadow, Midnight, and Chichi.
Their unconditional love saw me through.

# CONTENTS

# AUTHOR'S NOTE

Oh my God, this is the highest I've ever been in my life. I can barely hold on. Sweat is pouring out of me, my eyes sting like hell, and my gut is jumping. I'm completely soaked, my arms and legs flail, my head shakes, and my heart is thumping out of my chest. I am flying, and I love it. I desperately want this feeling to go on forever. I'm Steven Adler, the drummer for Guns N' Roses, and tonight we opened for the Rolling Stones. It's October 18, 1989, and after a brutal but amazing journey, this should be the happiest moment in my life. But as we explode into our last song, "Paradise City," I'm already terrified of leaving the stage and losing this incredible buzz. Just like the "pre-high" addicts get right after they score but before they use, I'm experiencing a devastating "pre-crash" and I fucking hate it. If only I could find some way to maintain this intense natural high, I would never need drugs, never want drugs again.

Guns N' Roses gets a standing ovation, but as the Stones take the stage with "Start Me Up," I'm already alone, tucked away in my trailer on the backstage lot with the door locked tight. Why? Because I'm the undisputed all-time booze-chugging, pill-gobbling, drug-shooting, Katrina-caliber fuckup. Throughout my wretched life there isn't a friend, family member, or fantastic opportunity that I haven't shoved into a blender and mutilated.

But people love train wrecks. They just can't look away from the ODs, lawsuits, prison terms, rehabs, reality shows, meltdowns, and more ODs. So before one or all of the above happens again, I want to set the record straight. And I'm finally sober enough and angry enough to do it right.

## ⊰ COMING CLEAN ⊱

While part of this comes from a deep desire to come clean with my family, friends, and fans, another part is fueled by an inner rage to represent. From Chuck Berry to Janis Joplin, from Hendrix to Cobain, many beloved, gifted musicians have had a lot of totally false, negative crap written about them. It turned them into bitter, reclusive artists and may have pushed some into an early grave. But I don't need the media to bury me; I'll do that on my own.

The bastards who write the lies about us do it because they think rock fans are gossip-starved, tabloid-trained morons who will swallow anything. They figure the more bullshit they pile up, the more fans will be eager to roll around in it. And they're always sure they can get away with the most outrageous lies because they know if we hire a bunch of lawyers to go after them, they'll just get more free publicity and rake in more cash. You've got to admire Carol Burnett, Kate Hudson, and others who brought suit, hung in there, and won judgments against these bloodsuckers.

The truth is I'm healthier and happier than I've been in twenty years and I refuse to be destroyed by all the negative news about Steven Adler. I've made it way too easy for these jerks to write me off as just another has-been junkie asshole.

And hey, I admit it. I am a has-been junkie asshole. But there's a lot more to this drummer boy. With the help of Dr. Drew, and a lot of other dedicated professionals, I've begun to live again and love my family, friends, and music again. I know I've let them down, but that's not going to stop me from trying to get back up and make things right.

## ⊰ MY GNR BROTHERS ⊱

Axl, Duff, Izzy, and Slash, I pray you'll respect my desire to go on the record and tell everyone what actually happened. My goal here is to dig deep and, to the best of my knowledge, tell the whole truth and nothing but.

Now, that's not to say those guys don't recall different things, or things differently. But when it comes to writing about my life as a rock musician, Axl, Duff, Izzy, and Slash will be the first to tell you that I've been my own worst enemy. And I'll be the first to agree. This isn't about laying blame, it's about accepting it. And in spite of all the fuckups I've had, the love is still there. *A lot of it*. I still love every one of those guys, and I hope they know it.

One of the things Slash writes in the closing pages of his memoir *Slash* is that he's truly happy that "Steven Adler is doing better." I got very emotional when I read that. Slash and I have been through *so much*. Since we were thirteen! The fact is Slash has had a lot to do with my seeking help and letting the light back in my life. Thanks, Slash!

There is still so much affection there, so much shared pain and joy. You can't ever take that away. Not from me and Slash. And not from me and Duff, Izzy, or Axl. The only way to make these pages matter to me and you and everyone who has loved or hated me over the past forty years is to make the whole truth the price of admission—and Adler's admitting everything.

# FOREWORD *for the* DUSES

> *Now it's a mighty long way down*
> *rock 'n' roll*
> *As your name gets hot, your*
> *heart grows cold . . .*
>
> —"ALL THE WAY FROM MEMPHIS,"
> MOTT THE HOOPLE

Those lyrics are from Ian Hunter, lead vocalist for one of my favorite bands, Mott the Hoople. And it kind of sums up what we went through with GNR. The bigger we got the more stuck-up and out of touch we became. Hunter also wrote one of the greatest books ever about life on the road, called *Diary of a Rock 'n' Roll Star*. It takes all the shine off the glamorous rock star image and puts it in its proper unfiltered light. It is a frank, many times joyless account of what rock 'n' roll is like from the inside looking out.

Hunter was determined to get it all down in his personal account of Mott's five-week American tour in November and December of 1972. It should be required reading for all kids before they start smoking cigs, skipping school, and jamming in garages. Hunter talks about Mott's equipment getting stolen, concerts being canceled, and fans being abusive. Believe me, fifteen years later, when GNR toured the world for eighteen months straight, not much had changed.

Ian leaves nothing out because he knows that's the only way to offer the story. If you're going to tell it, tell it all. I want to thank Ian, Mick, Overend, Phally, and Buffin for inspiring me to give my readers the truest, most unflinching account of rock 'n' roll since Ian penned his masterpiece. If I can get close to the honesty and guts on those pages, then this will be a great book. And I will owe it all to you guys. You are and forever will be the Dudes, the original lineup, the first and the best.

## ⊰ SORTING OUT THE MESS ⊱

Great rock music, whether it's Mott or Mötley, has helped me crawl out of a hole where I've been living a permanent nightmare. For two decades I've been haunted by a shady, drug-addled past that sucks any desire to face life right out of me. But in the past year, leading right up to November 2009, when I performed with Slash, Duff, and David Navarro in a sold-out show at the Palace in L.A., the music's inspired me more than ever. It's lifted my spirits and made me want to live again so I can create music with my band, Adler's Appetite. I want to get back together with the faithful companion that never betrayed me, my drums.

Now, understand that many of the interviews I granted during and after Guns N' Roses are a lot of garbage. I tended to treat them like a game, varying what I said to dick around with whoever was interviewing me and drinking heavily before and during them, because a lot of interviews were tedious and repetitive.

Being sober changes everything. The light is harsh at first, and there's a lot more I'd rather forget than remember. But I've fought hard for the opportunity to come clean here and that means everything to me. Although it's terrifying to revisit how things got so twisted, it's also the only true way to get my life back. So let's start this journey at the beginning, so we can understand how things began to unravel until they got so *fucked up*.

# My Appetite
## FOR DESTRUCTION

# TROUBLE *from the* START

### ⚔ CLEVELAND COLETTI ⚔

was born in Cleveland in 1965, during a time when my father had sunk to physically threatening and beating my mother. Things had really deteriorated between the two of them over the six months before I was born, and Mom was already plotting her escape from this monster by the time I arrived. I was named Michael after my biological father. Poor Mom probably battled a gag reflex every time she said my name. My older brother by three years was named Tony. This was to honor the Italian tradition of naming the first-born son after the paternal grandfather in the family. The second son gets either the maternal grandfather's name or, as in my case, the father's name.

I guess this goes on in other cultures, which is why Bobby Kennedy named his first son Joe, after his father, and his second son Bobby Jr. But the tradition doesn't fly with Jewish families, where you absolutely *do not* name your kids after anyone who is living. I'm sure my Jewish mother never confronted my Catholic dad with that fact, because she probably wasn't eager for another beating.

My pop, Mike Coletti, was sadly just an Italian gangster-wannabe with a bad gambling problem and a worse temper. He and my mother, Deanna, married very young, before they were fitted with brains. A short time after they wed, he became verbally abusive with her, and it just kept getting worse. In fact, the last time my parents were together was the day he beat the hell out of my mom and left her bloody and unconscious on the front lawn of my grandmother's house.

Now, I know I was too young to remember that day. And most doctors would probably agree that such behavior wouldn't leave any lasting psychological scars on a newborn. But Mom said that unlike my older brother, I used to cry all the time, day and night. Even that used to piss off my dad, who was too cheap to pay the $30-a-week child support ordered by the judge after they split up. We never saw Dad again. My older brother did a Web search for him recently, finding out that he passed away in 2004. I honestly believe that I sensed there was no love between my mom and dad right out of the womb.

## ONE STEP AHEAD OF HITLER

So Mom left Dad and now, twenty-four, with two little kids, realized she had nowhere to go. She was in desperate need of help. Her relationship with her mother was nonexistent, but with absolutely no alternative, she asked her parents for assistance.

My grandmother, "Big Lilly," as I knew her, came to America from Warsaw in late 1939. She arrived in the United States just *three days* before Hitler's armies invaded Poland. Big Lilly lost her entire family to Nazi butchers during the Holocaust. This experience forged her into a fiercely independent woman. Our Jewish heritage was her raison d'être. It formed the basis for everything she held sacred in life. My grandmother's faith ran so deeply, it was the foundation of her very existence. And Judaism was the one solid rock my grandmother and grandfather could stand on when everything else was threatened. If they ran out of money, if their little bakery business failed, if they were sick, cold, or hungry, they were still the Chosen People with God on their side. They had the Jewish faith and that would get them through anything.

My mom pretty much pissed on that whole belief system. *Screw the Jews, I'm marrying a capicola Catholic. I'm in love outside the faith, and to hell with everything you've tried to beat into me over the past twenty years.* When Mom did this, the family's rabbi interpreted it as the most vicious attack on everything the Jews stood for and everything they sacrificed during the Holocaust. Big Lilly thought this could not be her child, because no one she raised could be that irreverent, that disrespectful. Imagine how humiliating it was for my grandmother to face the other Jews in her neighborhood, particularly at synagogue.

So Big Lilly believed she had no choice but to disown her daughter. I am torn because my mother was very young and married out of love, which is often so blind. But I guess if love is blind then marriage is an eye-opener.

## ⊰ CLASH OF CULTURES ⊱

After such banishment, what could possibly send Mom crawling back to my grandmother? It's simple: she had no other choice. We were freezing and starving. We needed to eat, and we needed someone to clothe us and keep us warm.

Now, Grandma wasn't totally heartless, but before she agreed to help my mom, she made it clear that certain conditions must be met. First, my brother and I had to change our names to adhere to the strict Jewish custom that I mentioned that allows no newborns to be named after living people. So to please Big Lilly, my mother renamed us. I was now Steven, and my brother was Kenny.

Grandma Lilly's second condition was pretty radical: Mom had to give me up. I was to live with and be raised by Big Lilly and my grandpa "Stormin' Norman." I literally became their son and spent most of my childhood under their care. Mom couldn't believe her son was being stolen away from her by *her parents.* I remember my mother's constant sobbing during this time when she'd be allowed to visit me. With an innocent child's perception I'd be thinking, "Ma, what's wrong? Ain't you happy to see me?"

My mom was completely crushed. It wasn't that I was her favorite or anything like that, it was just that I was her darling towheaded son, and that was enough. Now, I don't know about you, but that rates right up there for all-time fuck-yous. It was payback time, and Big Lilly wanted to show my mom that Italians aren't the only masters of revenge.

## ⊰ WILD CHILD ⊱

At this point in my life, I was pretty much like the free spirit in that song by the Doors, "Wild Child."

*Not your mother's or your father's child*
*You're our child, screamin' wild . . .*

I was one wild, crazy, fucked-up kid, a born contrarian. Anything, and I mean *anything*, I was told to do, I would instantly do the opposite or just completely reject it.

My earliest memories are of my getting into trouble. I was kicked out of school during the first week. I gathered up and threw wooden blocks as hard as I could at the window. I still remember the sound it made. At any moment the glass could have shattered. I kept laughing at the way the other kids would wince at the sound. Fuck 'em.

As soon as the teacher stopped me from doing that, I tricked some kid into helping me get something out of a closet filled with board games and toys. As soon as he stepped in front of me, I backed off and slammed the door, locking him in.

He immediately had some severe claustrophobic episode. He started screaming at the top of his lungs and pounding on the door. To compound things, the teacher couldn't find the key to the door right away, and the whole class became freaked out listening to this kid lose his shit.

When the teacher tried to discipline me, I threw a temper tantrum and pushed her as hard as I could. It seemed like I was locked into this other world and every time teachers told me to do something, they threatened the universe I lived in, and I had to fight them with all my might to defend my world. How dare they be a menace to the galaxies I ruled?

To their credit, the school principal and the teachers believed I had a likable side but had some control issues. They put up with an awful lot for a little time, and then they expelled me from preschool.

## ⊰ SPOILED EGG ⊱

Regardless of my behavior, Big Lilly was determined to spoil me. She really did everything for her little bratty, impulsive grandson. But sometimes I'd go too far even for her and she would ship me downstairs to be with Kenny and Mom. This was when the whole family lived in the same claustrophobic complex in Cleveland. Of course it took only a day before I would get on Mom's nerves and she would send me to my room. Now, that's not going to work, Mom.

I would just open the window and scream out, "Grandma. Grandma!" We lived on the fifth floor and my grandma lived on the ninth. She'd come down, forgetting all about the terror I had been the previous day, and run to my defense. She always stuck up for me. She seemed to take pleasure in ordering my mom around and demanding I not be punished because I was "a very sweet boy."

She delighted in seeing Mom squirm. Mom was livid with the way Grandma Lilly and I would gang up on her, and there was really nothing she could do about it. When I'd get back upstairs, I could do no wrong until I wore Big Lilly down again.

One of my mom's older sisters lived in California. They used to keep in touch over the phone at least twice a week. Whenever they talked, my aunt would always tell my mom how great it was to live in Southern California. She would excitedly brag about the weather, the beautiful beaches, the ocean, canyons, and mountains. You could do anything at any time, because it was always sunny and warm, even in the winter.

Eventually this got my mom thinking about busting a move out of Cleveland, where, in all honesty, things couldn't have gotten much worse for her. One day, Mom started asking her sister about job openings, and my aunt was ready. She grabbed the paper and actually started to rattle off openings she had circled out of that day's classified ads.

Mom would usually tell us all about their chats after she got off the phone. When she was on the phone with her sister her voice would actually get sweeter and go up about an octave. She'd even speak faster and we could tell she was getting more and more excited with each call.

Finally, the chance for a fresh start and a new life overcame any fear or reservations she had. Moving was something she had been mulling over for months, and one day, when I was having dinner at her place, she just sat us down.

I'll never forget the look on her face. She took each of her boys by the hand and told us we were going on an adventure. We were going to visit her sister out in California, and maybe even stay out there if things went right.

The fact that she picked one of the coldest, windiest, wettest days of the winter to tell us certainly sealed our approval. Kenny and I were all for it. I never saw my mom so wired. She talked about our move nonstop. Maybe that was to hide how scared she was during this whole time. She would make lists, then make more lists, then tear them up, make some phone calls, and start a new list.

She read travel brochures on Southern California. Then she boxed up everything we owned, from her sewing kit to the salad bowl, and labeled it all with Magic Markers. The entire time she had this look in her eye, like a runaway train. God pity anyone who got in her way. That's probably why I never heard Big Lilly put up a fight for me when the time came to head out.

I could feel the excitement as the date neared. This golden opportunity to get a bad start behind her and begin again with a new home gave her boundless energy. She could have sprinted to L.A. So, at the lucky age of seven, we drove to California to get a fresh shot at life.

# 2

# GOING *to* CALIFORNIA

*Made up my mind to make a new start,*
*Going to California with an aching in my heart.*

—"GOING TO CALIFORNIA,"
LED ZEPPELIN

Mom found us a tiny apartment in North Hollywood. This began a second series of long phone conversations, but this time it was with her boyfriend from Cleveland, Melvin Adler. A month later Mel showed up with a huge suitcase and a big smile. Mel and the suitcase never left. Even though we were literally living on top of one another it somehow worked out so well that in 1973, Mom and Mel got hitched.

## ⊰ NEW ARRIVAL ⊱

In 1975, Mel and my mom became the proud parents of a baby boy, Jamie. Just before mom had Jamie, Mel believed it was time for us to officially become one big happy family. He spoke with Mom, then asked Kenny and me if he could adopt us. We were thrilled and had our surname legally changed to Adler.

Jamie lit up our world. I loved my little brother so much I decided that I was going to protect him. Next to his crib was a small couch where I slept every night with a switchblade in my hand.

Nobody was going to bring any harm to my little brother ever. I'd kill them if they tried. I'll never forget the flash of alarm in my mom's eyes when she spotted the knife, but just as she was about to explode, she caught herself and leaned over to kiss me gently on the head.

Somehow my parents knew this was just a phase and they never freaked out about the knife. Long after I stopped guarding Jamie, I continued babysitting him and even changed his diapers (well, just a couple of times . . .).

With Mel and my mom both working we were able to move into a bigger house in Canoga Park. Mel got a steady job, one he held until he became very ill in 1991, as a chief clerk for the Southern Pacific Railroad. Mom worked as a waitress at a restaurant called Two Guys from Italy. (What the hell is it with Mom and Italians?) Since most of Mom's family, her three sisters and a brother, had now settled in California, my grandparents soon followed suit and moved to Hollywood.

## ⊰ SIBLING DEATH MATCH ⊱

While I couldn't have been closer to Jamie, I had to share a bedroom with Kenny, and we could not get along. We really hated each other. We fought all the time. He was always taunting me. It may have just been a run-of-the-mill sibling rivalry but it soon got way out of hand. He would tease the hell out of me and push me to my limits. I'd put up with all I could take, then fight back as fiercely as possible.

He was a lot bigger than I was, so I often found myself on the losing end of our brawls. And it wasn't always a physical battle; many times it was mental torture too.

Like when Kenny had this paper route. He saved up enough to buy a cheap used TV. At night, he would turn the volume way up and position the set where I couldn't see it from my side of the room. He'd be laughing at *The Tonight Show* or whatever while I just lay awake unable to see the TV screen or get to sleep.

One time I got so furious with him that I smacked him with a tennis racket in the back of his head with all my might. He keeled forward in the bedroom like he had been shot. Good thing he collapsed on the bed. Kenny didn't move for like five minutes. He suffered a concussion and I really caught hell from Mom, who screamed at me for an hour.

## ⊰ OPPOSITES REPEL ⊱

We were complete opposites in every way. Kenny resembled our dad, olive skinned, dark haired, and heavyset. I was thin and light like Mom. We never went to the same school at the same time. It always worked out that when I'd be entering junior high or whatever, he'd be graduating. In class he was shy and introverted. I, however, was very outgoing. I made the class laugh and made friends easily, usually hanging out with older kids who were almost my brother's age. Kenny preferred to hide out in our room, reading comic books and watching TV. He was content to do that all the time.

Things hadn't changed much from my kindergarten days. I would get in trouble nearly every day. I was still getting in fights and talking back to teachers. My mom received calls from the faculty. Teachers, coaches, classmates, the custodian—I didn't take shit from anybody.

## ⊰ SUMMERTIME BLUES ⊱

Mom and Mel were constantly trying to figure out ways my loner brother Kenny could make new friends. So one summer they sent my brother and me to one of those Hebrew summer camps. Clear Creek I think was its name. I got there and just went nuts.

Poked around, made fast friends with everyone, then fast enemies. I was so bored by the end of the first week that I thought I was going to go insane. So I did.

I've always had an imagination that lets me visualize doing something or being something before even a hint of it happens. Sometimes it serves me well, like when I told Slash we were going to be huge rock stars, but most times it's just the forecaster of doom. My doom. Big doom.

At the end of the second week, when they had "family day," my parents proudly came to visit. They were expecting to hear fun stories about what a great time we were having. They were expecting the counselors to tell them what swell kids we were. They thought they were getting a lovely day out in nature.

They got something else. Mom and Mel sat there numbly as the counselors told them I had been running wild in the camp and had probably been the one who stole $300 from one of the camp counselors while she was in the shower.

Can you imagine the shock on my mom's face? One moment she's walking along this idyllic tree-shaded lane with Mel. She meets us at the lakefront, all adorable with little Sunfish sailboats bobbing in the background. She sits down to hear the camp counselor telling her that I'm a thief and a liar.

I had already denied it—they had nothing on me—and besides, I knew who did it.

They pulled a full-on search of the camp. Since I was the usual suspect when it came to evil and mayhem, they interrogated me about the missing cash. Three male teenage counselors held me down and forcibly searched me. Needless to say, a situation like that could easily provoke a lawsuit today. I was barely ten.

They didn't find the cash, and I was tempted to act outraged and demand my parents seek some kind of restitution, but in the end, this little demon that nests in my head received a sharp pang of guilt. That girl counselor looked like walking death. It must have been her life savings.

Yeah, I knew who did it. I did it. But honestly, it was just for kicks. I was so fucking bored by the second week, I wanted to spice things up. So I bought a shitload of candy with some of the money I stole but then, instead of stashing it, I gave it out to everyone. I know, brilliant.

Just when they were getting a little shaky over their accusations, I confessed and gave it back of my own free will, minus what I had blown on candy. It was like as soon as I thought I might actually get away with it, then I felt free to confess. I was more into taking it for the sheer thrill of it anyway.

So on "family day" my parents, much to their dismay, were treated to a request to take me home. My brother was allowed to stay another two weeks. They liked Kenny. He overate, never questioned anything, and kept largely to himself: the perfect, no-trouble zombie camper.

## ❧ BACK TO CIVILIZATION ❧

During the car ride home, we sat in icy silence. I couldn't have cared less. Camp wasn't my idea. It wasn't camping anyway. Camping is going to Yosemite and hiking up where no one can possibly find you, packing nothing but a PowerBar, canteen, and sleeping bag.

You sleep under the stars for a week. You eat roots and berries, spy on wildlife, and smell like ass before you hike out again. That's camping, which I also had no interest in doing.

The only thing I had any desire to do during that silent, unending, and tense drive back to Canoga Park was to get together with my two best friends, Ricardo and Jackie. They lived just down the street from me. Ricardo was a cheerful Hispanic dude. His mom made the absolute best salsa ever. Jackie was Asian, the mellowest wingman you could ever want to meet.

Jackie went to a local elementary school and Ricardo and I both went to Limerick Elementary. We also played in the junior football league, but on different teams. We were very competitive and had each built a reputation for being fast and tough. The season culminated in a game between our respective teams.

My team lost. To be honest, with the exception of a few dedicated players, we sucked. I think we lost every game we played that year. The league was for ages eight to twelve, and I swear, no one on our team was older than ten. The apathy at school was rampant. The older kids were either too cool or too spineless to step up and play.

I used to think we were always pitted against impossible odds. All the other teams had older kids and they were much bigger than us. We developed a humiliating reputation within the league. Instead of the Eagles, kids called us the Bad News Birds.

On my football team, I was the starting running back and kick returner. I even won a trophy my junior year for Most Valuable Player. The coach used to put me in and keep me in for the entire game. He told my parents I was the best player on the team, but it didn't matter. Even I couldn't do anything to end our losing streak.

## ⊰ SEX AND SPORTS ⊱

'll never forget this one time during practice. There was a gorgeous cheerleader hanging out on the sidelines. I couldn't help but notice her. I was just eleven years old, but I was already developing a healthy appetite for the opposite sex.

But when I'd talk to my friends about how nice so and so's ass was they'd just look at me like I was a freak. "Asses are gross." "An ass is an ass." It's like their cocks hadn't kicked in yet.

Anyway, she was way out of my league, sweet sixteen with long blond hair and these amazing pouty lips, like a crushed rosebud, all full, round, and soft and begging to be kissed. I *just had* to get her attention.

We were having a team scrimmage and when I got in the huddle, I was so amped to impress this babe that I threatened the quarterback. I told him he better give me the ball or I'd smash his face in when we got in the locker room. I swear something snapped inside, and my whole world came down to impressing this cheerleader. Every time I got the handoff I ran like a possessed demon for a touchdown. The coaches were stunned. I scored like five TDs in a row.

I don't know why I'm wired this way, but there are very few things in life that really light me up. And nothing focuses me or gets me going like chasing tail. Money, fame, status, power . . . nothing comes close to the pursuit of pussy. It gives me an intensity that brings out the fiercest side of my competitive spirit.

When I was with the band I *had* to score the best snapper after a concert. I loved parading around backstage and at the after parties with the pick of the litter. So whether it's trying to score by making touchdowns or playing in a band, I love the ladies. Primo poon: accept no substitutes.

## ⊰ BAD MOVES ⊱

After mustering a big smile, I went over to the girl after practice and said something that I thought was cool enough to get a kiss off her. But she gave me such a look. Ouch! Then she just turned away as she muttered something about waiting for her linebacker boyfriend to come over to her after practice. I was so crushed.

As the rest of the guys filed off they looked pissed at me for being such a showoff. I remember shaking my head and letting out this huge sigh. I couldn't believe what an asshole I had been, and all for nothing.

Just as I started walking away from the bleachers, she turned back toward me and gave me a little smile, saying, "What's your name?" I got this big lump in my throat and croaked out, "Steven." She repeated my name, nice and low, and believe me, that made it all worth it. To this day, I can still hear the way she said my name.

## ⇥ HANGING WITH MY BLOODS ⇤

Outside of school sports, Jackie, Ricardo, and I spent every minute together. Ricardo was going out with this cute little blond girl at the time, but he always put us first. Nothing was more important than the bond between us. At least that's what I thought until I received my first lesson in the politics of friendship.

Ricardo and I found some oranges in an abandoned lunch bag at the playground, and we started throwing them back and forth at each other. One orange started to break up from hitting the ground too much. I remember throwing it way up in the air toward Ricardo, who was like thirty feet away. All of a sudden, his little blond babe starts skipping right over to him, and *blam!* The orange came down and just *nailed* her on the head.

She was screaming, covered in orange. Ricardo freaked and started chasing me all over the field. "You're dead!" he yelled. I tried to sprint away but he caught me and got on top of me. I was helpless. He had my arms pinned with his knees. I thought he was gonna start punching me in the face, or at least spit on me, but he didn't. I guess he realized that she had already run home and it really was just an accident. But he was really pissed at me. Over a girl . . . a *girl!*

## ⇥ THE SEVENTIES ⇤

The seventies were a magical time, especially for a kid my age. It was the perfect decade for growing up. I remember seeing Kiss records in stores, before I had even heard their music. I thought they looked so cool. And I loved *Charlie's Angels.* Jaclyn Smith was my favorite. Of course, *Happy Days* was a big show for me. I wanted to be like Fonzie.

I collected Stop N Go gas station mini NFL football helmets, which you could only get by purchasing Stop N Go's inferior version of a Slurpee. I had to have them all, and as quickly as possible, which meant plenty of brain freezes!

I wore tight Sassoon pants, corduroys, or Levi's. Bell-bottoms were at their peak of popularity, and *everyone* had to have Vans tennis shoes. The *cool* thing to do was to have them custom-made. You would have to wait a few weeks, but it was worth it. Just *having* Vans was cool, really. But they cost about $40 . . . Hey, Grandma!

White moccasins were another hip thing to wear. They were available only in the leather stores at the farmers market. My grandparents would take me to get them. The employees would see us coming in and they would take them down for me, knowing exactly why I was there. They cost like thirty bucks, but my grandma never had a problem buying me new ones once the old pair wore out.

I also loved playing with yo-yos. This was when I was ten years old. They're not very popular today, but back then they were pretty common for kids, a must-have toy that was advertised all over TV. I became a pro. Every week, we had these yo-yo contests at our local 7-Eleven store, sponsored by Duncan yo-yos, right in the parking lot. Ricardo and Jackie weren't into it as much, but they would still come along with me on their bikes. I could do every popular move, and I even made up a few of my own. I was in the running to win every time.

As a prize, they'd give me a nice yo-yo, like a glow-in-the-dark one. I won at least ten of those things. I rocked at that and always came home excited. In my mind, I was a stud who could be the best at anything I put my mind to, particularly if the girls thought it was cool.

## ⊰ THE BEGINNING OF THE DRUG SCENE ⊱

As we got older, Ricardo, Jackie, and I became aware of the drug culture that so many other kids were getting into. It was 1977 and a new phrase that epitomized the attitude of the time swept the nation in an Ian Dury and the Blockheads song called "Sex and Drugs and Rock and Roll." It was in the air and on my mind.

My friends and I were very curious about drugs, and it wasn't long before the fateful day arrived. Jackie and Ricardo must have experimented before me because it all started this way: one day we were all hanging out and Jackie asked if I wanted to get high. Just like that—out of the blue. I knew exactly what he meant. Ricardo had a makeshift pipe he had made out of tinfoil. We walked over to Winnetka Park and sparked up.

We were sitting in this deserted dugout and I remember the first wave of cannabis hitting my medulla. It was subtle at first, then *whoa*. The sounds were what tripped me out most. Ricardo's voice sounded so different, and the colors behind him, different shades of green on the trees, shadows slipping in and out of the backstop. I couldn't help but feel that was the greatest, most profound fucking moment. I discovered what I thought was heaven on earth that day. I thought that I had found a connection with God. Then I just started cracking up. There was a Taco Bell just across the street from the park, but I couldn't order anything because I couldn't stop laughing.

At the wise old age of eleven, my friends and I entered a new chapter in our lives, and our daily activities changed completely. While we used to bike around, or play catch or whatever, we now smoked weed almost exclusively. We never drank. Jackie began selling weed, so we would always have a supply. After school, we'd be at one of our houses, just watching TV and getting high. Our parents worked, so it was a carefree time. Those were the days; it was summer, no school, just hanging out. Good times with good friends.

It was also at this time that we had a conversation that I consider to be the most prophetic moment of my life. We were eleven years old, hanging out in Ricardo's backyard, sitting on his dad's tractor. It was another gorgeous sunny day in the Southland and we had just finished up a nice fatty. All of a sudden Ricardo goes, "I want to be in construction like my dad." Then Jackie says, "I want to be a mechanic like my dad." I looked at both of them and all I could think about was Steve Tyler rocking out, shouting, "Dream on! Dream on!" so I blurted out: "Well, I'm gonna be a rock star!"

## THE CRUX OF THE BISCUIT

So let me digress for a moment, because this is really important. And while it might not be as sensational as Axl's chaps or Izzy shooting up or Slash fucking strippers, to me this is more important. It's revelatory and from the heart.

Whether you want to be a rock star, or play in the Super Bowl, or go to Harvard, you just got to say it out loud and believe it. That's it. But you've got to have 100 percent unwavering faith in what you're saying. It's that simple. I did it, the guy who's helping me write this book did it, and we both know other people who have done it too. Rock star, Harvard, Super Bowl—believe it and you will be it.

## FIRST GIG

When I was twelve, I got my first part-time job at the Pioneer Chicken fast-food restaurant, which was right by where my mom worked at Brent's Delicatessen. I cooked the chicken and I cleaned floors, whatever they needed me to do. I was at that stage where I'd rather have been making some money than going to school. By now, I was in seventh grade at Sutter Junior High School. I hated it, and I wanted out. The first time I ditched, I remember walking out of the school, shaking because I was so scared I was going to get caught. I walked out the gate and crossed the street, and waited for the inevitable shout from a teacher, but nothing happened. That serious lack of school supervision definitely encouraged me to ditch school every day.

Every morning I'd get on the bus, which only cost like forty-five cents back then; buy a transfer, which was like thirty-five cents; and ride up Winnetka, which was where the school was. But I wouldn't get off at the school; I would just keep going. The bus would continue down Ventura Boulevard, and I'd get off at the hill in front of Universal Studios.

I used to hang out in the area where people would come out and shop after riding the tram. They had Frankenstein walking around, and people dressed as cowboys and Indians performing the stunt shows. One time I arrived early and met the villain of their little live performances. He sported a sinister, thin mustache and an all-black outfit and whip. He'd shoot the good guys off their horses and stuff. His name was Lance Reamer, a man in his fifties. He used to go to the restaurant where my mom worked, so I introduced myself as her son and said that she'd told me about him.

Lance would let me hang out backstage, and I loved it. It had a really cool vibe. Lance never asked me why I wasn't in school and actually became a good friend. That was when I realized I wanted to be a stuntman. That lasted about a month.

## ⤳ CUTTING CLASS, SCORING ASS ⤝

I was supposed to be in school, and while cutting out one day, I met another kid ditching classes from another local middle school. His name was Josh. He had shaggy dirty-blond hair and wore a brown leather-fringe jacket. We were just walking around the neighborhood, during an unusually cold drizzly afternoon, when we ran into a pair of twelve-year-old girls, apparently cutting class too.

Josh had a pack of cigarettes and shared them with us. We all hit it off quickly. One of the girls had the Marcia Brady look: long, straight blond hair. The other had the lengthy, curled-back bangs that Farrah Fawcett made so popular at that time. As we talked and laughed, we made our way over to a construction site. We entered one of the half-built houses and looked around. There was just the stud framing supporting some Sheetrock walls and a ton of that cheap multicolored foam padding you see when you rip up wall-to-wall carpeting.

The place had a nice homey feel to it, and it was obvious other kids had hung out there. Someone had tipped over one of those big wooden spools for a makeshift table and dragged over a bunch of cinder blocks for seats. There was even a Led Zeppelin mural spray-painted on one of the walls. It was the four symbols for Page, Plant, Bonham, and Jones included on the inner sleeve of the fourth album. On another wall someone had spray-painted a crappy version of the Blue Öyster Cult logo. Once inside, we casually walked around as if we were thinking about buying the place, then paired off into separate areas. Josh disappeared first with Farrah, leaving me with Marcia Marcia Marcia.

Marcia was the prettiest young thing. The lipstick and light blue eye shadow made her look a bit like a windup doll, and I began to fantasize about what lay ahead. As she peeled off with me to find a quiet place, I could hear her steady breathing turn fluttery, expectant. I was suddenly, acutely aware of a sweetly fresh fragrance wafting off her body. I inhaled deeply, feeding off the seductive bouquet. It filled every pore in my body and made me so hot. Flush with excitement, I ducked into one of the smaller rooms, then turned to face her. Without hesitation she collapsed into me, surrendering completely. Her motion caused her long blond hair to fall forward against my face and shoulders. I thought I was going to lose it right there.

In one thumping heartbeat we were stretched out on a partially unrolled section of carpet padding in a half-built house, moaning, groping, sucking face, and praying it would never stop. I moved in closer to Marcia and she eagerly embraced me. I kissed her again and again, and she returned each kiss fully. Without thinking, I just seemed to know what to do next. In fact I swear I never had a single thought during the entire time. It was strictly, gloriously physical. After kissing our lips raw, we took a breath and I sat up, completely confident as I undid her belt. As I yanked on her jeans I heard the fateful word: "Wait . . ."

"Oh shit!" But she just smiled and asked me not to pull down so hard. She just wanted to help me get her pants off. She kicked off her sneakers to reveal these brightly colored rainbow-striped socks—so cute! Now her kisses were bolder, more urgent. As her thin white legs wrapped around mine, the full scent of her body hit me and I felt ready to burst.

From then on everything accelerated. She was so eager, so wet, that we were instantly one. The taste of her skin was both salty and sweet. We moved awkwardly at first, then got into a primal rhythm that couldn't have lasted more than a dozen seconds because I just exploded. I made this weird yelping noise that didn't even sound human.

This was my first time, and I think it was hers too. Afterward, she seemed to be equal parts nervous and excited, but all I remember was this enormous relief that there was no blood, and to a lesser extent, that I had performed. Yes, it was "mission accomplished" (though I had no aircraft carrier to string my message across). Honestly though, I was never nervous. Maybe that's because it was all over in a magic minute. I lost my virginity to my Marcia, a girl whose real name I'll never know. To this day, chilly, damp afternoons bring me right back to my first time.

## SHARING MY FRIENDS

At home, my older brother was withdrawing further, becoming a genuine recluse. My mom actually talked to the friends that I had, who were mostly older kids. She asked them not to hang out with me as much and spend some of their time with my brother. After all, they were my brother's age and they went to the same school. I don't know why, but this didn't really bother me. I never had trouble making friends, and Kenny could have them all. I'd just go out and make more.

They actually did my mom the favor and started hanging out with him. It worked out, because my brother had some new friends and I still got to hang out with them. Mom was fine with my being out later on school nights because I was with Kenny. In fact, my not being possessive with my friends ended up paying huge dividends because a few houses down, some of them were in a garage band.

When Kenny and I dropped in, my eyes bugged. I thought they were the coolest kids in the world. They were all tuning up their instruments and playing with their amps, and they all had long hair. I really looked up to all of these guys and treated them like they were gods.

They played Rush, Frampton, REO Speedwagon, Humble Pie, all the big rock tunes of the day. Hell, I knew them all; I was a rocker. The drummer had a blue translucent drum set and I remember he would roll joints on the snare drum. They had two guitars, a bass, and drums. It was so fucking loud. It was the first time I felt the actual, physical crush of live rock music. I fell in love with it instantly.

During one of their rehearsals, I did a beer bong with them. It was the first time for me. For those of you who grew up in a convent, a beer bong is a funnel attached to a tube. You put the tube in your mouth, and they pour an entire beer in the funnel. This forces you to swallow it in one gulp.

I was a wild man, determined to impress them all. I chugged six Olde English 800s in a row. I can still remember them cheering me on: *Ad-ler. Ad-ler. Ad-ler!* I felt like I really belonged. I smiled and laughed, proud I was able to entertain guys I so admired. When it was time for me to head home for dinner, I grabbed my bike. I had a yellow ten-speed at the time. I hopped on, pushed off, and watched in horror as my foot completely missed the pedal. I fell right off doing a header into the lawn. I was so wasted.

Two of the guys heard me fall and rushed over. "Where's your brother?" Kenny had wisely taken off, probably after my first beer. So these guys helped me up and walked me and my bike back home. When we got to my house, they set my bike down, rang the doorbell, and ran off. To this day, I wish I had run off with them.

# 3

# GROWING UP

About a month later, my mother was having a Tupperware party. I came in shit-faced out of my mind and said real casual, "Hey, Ma." There were about a dozen sweet old ladies right there in the dining room. I smiled; they smiled back. I thought I was so slick, fooling them all.

But then I felt a little shaky. I grabbed the back of the couch to stop from keeling over. Suddenly I power-booted all over the place, right in front of them. Technicolor Yawn, the Big Spit, Ralph-a-Roni. It's amazing how much more you can puke up after you think you're finished. I saw the nub of a hot dog in there that I'm pretty sure I'd eaten two days ago.

These women went from shifting uneasily in their seats to wanting the fuck out of there. They were afraid I was going to fill up the room with vomit and they'd drown. They certainly had a right to panic, because it felt like I yakked for a half hour.

Judging by the frozen look of horror on their faces, I was the girl in *The Exorcist*. Hell, I may as well have been Satan himself. Then

Mel flew into the room. He completely lost it, swearing up and down, screaming like a drill sergeant that I was grounded for life.

So there I was, sick as a dog. They sent me straight up to bed. I had been skipping school every day for about a month by that point, and as luck would have it, the day after I ruined my mom's party (and sofa), the school called my home.

That's when my dad, Mel, who was still livid about the barf, blew a gasket and screamed at my mom: "That's it. It's either him or me." I found out her choice the next day when I came home from the 7-Eleven where I had snuck out for a Slurpee. It was the only thing I felt I could coax down at the time; I was so hungover and dehydrated.

I turned the corner back to my house and . . . what the fuck? Mel had taken all my stuff and dumped it on the sidewalk. My clothes, a football, a couple of eight-tracks, and whatever else I owned was now all outside. I went in to ask what was going on, and all I remember was my dad and me getting into a big argument and his chasing me around the house.

But although I was screaming back at Mel and telling him what a totally unreasonable, heartless jerk he was, I knew deep down this day was coming. And while I've picked up lots of sympathy points over the years by telling people my parents kicked me out on the streets when I was only eleven, I had probably pushed them past the limit more quickly than your average juvenile terror.

So I guess it was time for me to leave. That was it. They paid the bills, they called the shots, so I was out of there. My grandpop, Stormin' Norman, came and picked me up. He helped me put my junk in his car.

### ⊰ OUT ON MY ASS ⊱

My grandparents had a small two-bedroom apartment in Hollywood about twenty miles from my parents' place in Canoga Park. After their children grew up, they each preferred having their own bedroom. Since there wasn't a third bedroom for me, my room was the bathroom. I kept my clothes in there, and a little clock radio, and that was it. The record player was in my grandfather's room, so I kind of adopted it as my own. Grandpa worked at a bakery, and he would leave for work early in the evening, so I'd always be hanging out in Stormin' Norman's room listening to my records.

It was hilarious. My seventy-year-old grandfather slept in this room that I had covered with pictures from teen magazines, mostly rock stars like Aerosmith, Boston, a lot of Kiss, Bay City Rollers, even Shaun Cassidy and Leif Garrett. I dreamed of being a teen idol. Grandpa never complained about my decorating, and Big Lilly never complained about the noise. I would sleep on their sofa bed in the living room.

Grandpa would come home from the bakery at five in the morning and he'd have a shot of whiskey. I'd be drowsy but I'd always ask, "Hey, Grandpa, how are you?"

"Just fine. Just fine. Hey, Stevie, wanna snort?" He called a shot a "snort."

I'd politely decline: "Nah, Grandpa." He'd always offer me a shot, do one himself, then go in his room. He'd close the door and go to sleep. Snort.

## ⊰ MY FIRST CONCERT ⊱

In June '78, I saw my first concert. My cousin Karen won tickets over the phone from the radio station 93 KHJ. She called me up and asked, "Stevie, how would you like to see Kiss at Magic Mountain?" My jaw dropped. She knew Kiss was my favorite band. I told her how much I'd love to go. The next day she picked me up, and we drove out to the amusement park. We were totally into it when we noticed there were camera crews setting up. They were filming Kiss for what would become their cult-classic film *Kiss Meets the Phantom of the Park*. I was a part of Kisstory!

That night I saw a lot and learned a lot about rock music. The most important thing I took away from the concert was an appreciation for how much the studio version of a song could take on a life of its own when it was performed live. It was the same song, same lyrics, same chord progression, but it was totally different, having a unique and often superior energy all its own.

When I returned home, there was no question. After the garage band experience and now seeing Kiss in all their glory, rock 'n' roll was for me! I begged Big Lilly for a guitar. She surprisingly put up little fuss and within a week, I had a Sears department store guitar and amp combo. Unfortunately, I didn't spend much free time practicing my new guitar.

Now that I was back in Hollywood, Big Lilly insisted I enroll at Bancroft Junior High. After classes, I would join all the local teenagers and ride bikes and skateboards at Laurel Elementary School, where they had all these cool ramps and dips and big embankments that the riders would race off.

One day I skateboarded up this ramp, taking off at a pretty good speed. While in the air, I attempted a 180 and messed up big-time. My head *slammed* down against the pavement—skateboarding helmets were not even in existence back then—and it felt like a bomb went off in my head. I was in such pain, bordering on passing out. As these two kids were walking by, they saw me hit the ground. They ran over to see if I was okay. One of them asked, "Dude, are you all right?"

I remember half rolling up to face them, holding my head. "Yeah . . ."

"Well, let's try to be a little more careful next time." They kind of snickered as they walked off. I went home with a huge lump on my head. This is probably when all the brain damage started.

## ⚔ SLASH AND I MEET ⚔

A couple of days later I was raising my own special brand of hell in class. My history teacher happened to have an apartment in the same building where I lived with Big Lilly. I tortured this poor woman relentlessly. See, I knew that she wouldn't want to get my grandmother all upset by telling her bad stuff about me, so I would take full advantage of that situation and fuck with her all the time.

One day, I had her chasing me around the classroom. I was crazy, running around the desks and stuff. She chased me into an adjoining classroom and the teacher in that room was in some other kid's face, pointing and shouting. "You're a loser, a bum, that's all you're ever going to be!" I thought that was amusing. "Hey, another kid who gets teachers to lose it." We looked at each other and just kind of smirked.

When the school day was over, we sought each other out. He said he recognized me from a few days earlier. I didn't make the connection at first, but here was the kid who asked me if I was all right when I smashed my head.

His name was Saul Hudson. And from the first fart, we got along great. That very day we started hanging out, climbing to the third-floor railing on the south end of the school. We were about six inches apart from each other, standing on one foot singing "King Tut," which was the happening Steve Martin song. We were laughing and having a good time until one of the teachers came out, yelling, "Jesus Christ, get the hell off there!" So we jumped and ran off. I think we bothered to show up at school for maybe another week after that, but then we started ditchin' every day.

We had a routine. First, we'd go to school just in time for roll call in homeroom. After we were marked as present we would leave. Sometimes we'd come back for lunch. We would sit at a table with this cool kid who had a huge head of big blond curls.

His name was Michael Balzary, aka "Mike B the Flea," aka "Flea." He played trumpet for the school band and later went on to form the Red Hot Chili Peppers. Turns out, he lived down the street from me. Most evenings in the fall, Saul, me, and a big group of kids would play football on my grandma's street. She would sit out on the front porch and take it all in. Most times, Flea would break out his trumpet and play for her. She thought he was a "little angel."

aul was the coolest, smoothest guy. He had already been scoring with chicks and had a definite air of confidence about him. Saul had many talents. We'd go back to the schoolyard where we first met and ride bikes. He was awesome on a bike, the best BMX-er around, and would even enter in competitions. Me, I had a Huffy that I hated. They were a joke to kids who took their riding seriously.

Saul lived with his grandmother at Sweetzer and Santa Monica. I lived with my grandmother at Hayworth and Santa Monica, five blocks away. The first time he came over to my grandmother's house, I showed him my guitar and little amp that my mom had bought me. I knew two chords and two scales.

Some kid had shown me those along with the main riffs of Chicago's "25 or 6 to 4" and Queen's "Tie Your Mother Down." I loved Kiss, so I put *Alive!* on my record player, turned the volume all the way up, and started rocking out, doing my best Ace Frehley impersonation. He instantly fell in love with the blistering noise coming out of the amplifier.

That's all it was really, just noise. That same week, I gave him the guitar to work with, and I decided to be a singer. I bought a mike stand and a cheap microphone. We would sit in his grandma's stairwell, and he would play guitar and write lyrics.

Steven Tyler of Aerosmith had all these scarves on his microphone stand, so I got these bandannas at an army-navy store at Vine and Santa Monica and put them on mine. These little kids who lived next door to me thought I was Leif Garrett. I had hair just like him. When I would run into them outside, they'd ask me for my autograph. Sadly, it was a short career, because while I really tried to sing, I just couldn't. I'd always sing to the radio and eventually I realized I wasn't any good. I also came to the conclusion that guitars weren't for me either. Drums gave me a primal thrill that a guitar could never touch. And I had been banging on Tupperware since I was two, so it was the drums for me.

I thought drums looked so cool. My first drum set was comprised of books stacked in position, and for my first drumsticks I used the bottom part of wooden hangers. Around the same time, Saul's grandma bought him a better guitar than the one that I gave him.

Again we would sit in his grandma's stairwell and jam for hours. Soon we became inseparable, like four people engaged in a lively conversation: Saul, his guitar, me, and my drums. I remember reading that when Clapton met Duane Allman after an Allman Brothers concert in Florida, their guitars talked to each other for hours. It was and always will be the music. The music bonded us for life.

## ⤙ THE FIRST JAM ⤚

The first time we actually performed a song, we were at some black guy's house, a kid we met at the Granada Hills music store. He played the drums, his brother played guitar, and they performed the Beatles' "Day Tripper" for us. I closely watched him play, sort of learning from his movements.

They let Saul and me have a go at it. For the first time I actually played on a real drum set. From watching the guys play, we mimicked what they did and played as best we could. We made some kind of whacked-out sound. But for kids who had no idea what they were doing, the seeds were there. We were born rockers.

Although Saul had just started playing guitar, he was really amazing from the very beginning. Saul was also writing kick-ass original songs, cool ones that had great hooks. In comparison, I wasn't very good on the drums. That wasn't going to stop us though. After that jam, Saul and I were all psyched to come up with a name for our band. After kicking a few around, one of us just blurted out "Road Crew." It's like when Robert Plant wrote the lyrics to "Stairway to Heaven," and he said that two-thirds of that song just poured out of him in like twenty minutes. He felt like he was channeling more than writing, and that's how we felt with "Road Crew." It was just sitting there in the cosmic realm, waiting for one of us to pluck it out.

God, we loved that name. It summed up our warrior attitude about bringing great rock 'n' roll to the masses. That was going to be our thing. Taking music from the streets to the streets. And when you think about our later success with GNR, that's what helped us click with our fans. They immediately recognized that this was *their* music, their own street brand of rock 'n' roll.

## ⊰ BLOOD BROTHERS ⊱

I had a skateboard and Saul had a bike. I would shred over to his house or he would ride his bike to mine. I was pretty much with Saul every day now, all the time. The music was our sacred bond, which is why I had drifted away from Ricardo and Jackie. I still loved those guys, but I loved rock 'n' roll more. Saul and I were slaves to the beat, hanging out on the stairway at his grandma's apartment building, writing music and lyrics. We were such good friends, so close, even doubling up to make it with chicks.

When we were both fourteen we said, "Let's do that blood brother thing." We got a knife, slit our hands, pressed them together, and said, "We're going to make it in a rock band and we're going to be huge." That promise formed an unbreakable bond between us. After that we felt we had sealed our success as a team.

## ⊰ STONER GIRL ⊱

Saul and I believed in sharing. Everything I had, he wanted. *Ha!* One time I was running over to his place when I passed by this guy and girl who were sitting out on their porch. The guy was rolling a joint, and the girl asked if I wanted to smoke it with them.

The girl introduced herself as Kerry. She was so beautiful. She had dark hair, big lips, blue eyes, and was extremely pretty, very exotic. I don't even remember the guy's name, but he was her roommate. This girl was twenty-three years old, and we just started hanging out every day while I was on my way to Saul's. I was only fourteen, but it didn't matter. After about two weeks of seeing her every day, I decided I was going to try to fuck her.

One night we just started making out, and I was so into her. I mean, right when my cock would touch her hot, wet pussy, I would pop. Three times it went like that. I just couldn't hold out. I had only fucked my little windup Marcia by then, and this older woman really rocked my world. My balls actually ached. And I was hard again in like ten minutes. Afterward, I went over to Saul's house and told him about it, all the facts in minute detail. Saul just got this odd, pinched-up look on his face, then disappeared for about fifteen minutes.

When he got back, I asked him where he was. He said he forgot to take out a load of garbage for his mom. I started laughing. "Yeah, sure. You had to get rid of a load all right." Saul got mad. He flipped his guitar pick at me. Stuff like that could really set him off. Then the next day, just to make peace, I introduced him to Kerry. Within just a few visits, he managed to score with her too. But unlike me, he knew what he was doing.

# 4
# LEARNING *the* HARD WAY

aul's main squeeze at this time was a beautiful young blond girl we'll call Melissa. We used to go over to her house all the time. I could just walk out the back of Grandma's house and I'd be on Melissa's street.

I learned how to roll joints from Melissa's mom, Carrie, who had a big bowl of pot and papers ready for me when I visited. One afternoon we were the only ones in the house. She was a very pretty lady, about thirty-six years old. She was twenty-two years older than me but as we were getting stoned I thought, "Y'know, I'm horny, what the fuck." I went for it. I started making out with Carrie, and she started moaning softly. I gave her a big wicked smile, then pulled out my cock and stuck it in her face. She started sucking on it, and it was incredible, a real blow job from an experienced lover. Although I was still learning, I felt like the hero of eighth graders everywhere. I wanted to get up on the roof and scream it out to the whole world. "I'm Steven Adler, the fuckmeister of grade-schoolers everywhere. Kneel before me!"

Without ever discussing it, we agreed it would be our little secret, once and done. There was no awkwardness between us, and the proof was that we remained good friends after our horizontal tango. It was natural, it was fantastic, and neither of us regretted a thing about it.

## ⇥ THE THREE OF US ⇤

But when it came to hanging out, it was just Melissa, Saul, and me. The three of us were inseparable for nearly four years. If you saw one of us, the other two weren't far away. I wouldn't come home for days. I slept over at Saul's grandma's house sometimes and at his mom Ola's too. Ola was incredibly talented, a lovely, artistic black woman who lived near Olympic and Crescent Heights.

Melissa had a friend, Michelle Young, who became a part of our tightly knit little group. She was a thin brunette with attitude. Michelle would be immortalized in the song "My Michelle" years later.

I remember the first time we went over to her pad, I was surprised by the stacks upon stacks of pornographic videotapes. She said her dad made his living as a producer in the porn industry and asked if I wanted to watch one.

I'd never seen one, so why not? This one took place in a jail cell, which was actually a really terrible set. These guys were getting their dicks sucked through the bars of this woman's cell. I was a horny young kid and found it difficult to hide my boner. But it was so damned cheesy, I had to laugh.

Every so often, we would visit Saul's dad, Tony Hudson. His dad hung out with a lot of rock 'n' roll people. He was a well-known album cover designer who did art for sixties folksingers like Joni Mitchell. I remember he took us to a couple of parties up in Laurel Canyon. It was so cool, a beautiful hippie house, everyone smoking pot, munching mushrooms, and it was the first time we ever tried wine coolers. The wine coolers were too lightweight for us. But everything else was just right. It was just nonstop sex, drugs, and rock 'n' roll.

## ⊰ WEED VERSUS WINE ⊱

Saul didn't like smoking weed as much as I did, so we would mostly drink when we hung out together. One week we'd have peppermint schnapps, another we'd switch to Jack Daniel's, and then another it would be vodka. Every week it was something different, depending on whatever struck our fancy (or what we found in the cupboard).

My grandpa always had a couple of bottles in the liquor cabinet, so we'd skim a little from there, or we would have someone buy a bottle for us. I'm pretty sure Melissa's mom would've gotten it for us if we really needed it.

I didn't smoke cigarettes at the time, but Saul loved his cowboy killers, Marlboro Reds in the hard pack. He'd always be like, "C'mon, smoke with me. I don't have anyone to smoke with." So without giving it much thought, I started smoking too.

We would roam anywhere and everywhere. At first he would ride his bike, and I'd hitch my skateboard behind him. Then we decided that since all the people we associated with were older than us, it was time to ditch the bike and the skateboard and just walk instead. Back and forth between all the popular Hollywood clubs: the Starwood, the Whisky, Gazzarri's, the Roxy, and our favorite, the Rainbow Bar and Grill. *Everybody* ended up at the Rainbow after the nightly club crawls. Saul and I had so much fun there.

Before going in, we loved to get primed by drinking in the parking lot of a nearby bank. One night we were pouring 151 Bacardi in the cap, lighting the rum on fire, and downing the mini shot. After a while, we were pretty toasted, and as Saul did his umpteenth shot, he missed his mouth completely and spilled a purplish-blue fireball onto his chin and cheek. All of a sudden the left side of his face lit up like he was the Human Torch. Saul didn't immediately realize what he had done and just looked at me like he was confused. I was shocked shitless but instinctively reached out to smother the flames with my bare hands before it did any real harm. Booze would definitely mete out its fair share of damage to us over the next decade, but not that night. Saul got away with a nice healthy glow on his face, and I didn't notice any burn marks on my hands. I'm sure we had both forgotten about it by the time we ordered our first round at the Rainbow.

Our afternoon strolls covered much of the same turf. We would also cruise up Santa Monica Boulevard, then head north past Barney's Beanery, a great bar where you could shoot pool, play foosball, and order some great chow. It was also where Jim Morrison wrote songs for the Doors and Janis Joplin hung out when she was in L.A. In fact the artist R. Crumb immortalized Barney's when he drew it on the *Cheap Thrills* album cover for Janis and her band, Big Brother and the Holding Company.

Saul and I would then head down Sunset to Tower Records, check out the scene there, and then wander up to Hollywood Boulevard. Tower isn't there anymore, and it makes me sad every time I drive by the old brick building. Some things, especially record stores, should never change.

Flipping through Tower Records' racks, from Aerosmith to the Who, always set me to dreaming about putting a rock band together, making out with our groupies, and traveling around the world. That's all Saul and I would talk about. He'd often have his acoustic guitar with him, idly strumming away.

I remember walking out of the store just as Benjamin Orr, the bass player from the Cars, drove by in his Rolls-Royce. He was the coolest-looking dude on earth that day. He had the top down, music cranked, and a beautiful girl with him. He just *looked* like a rock star. You could tell when you saw a rock star back then. They really stood out. And I just knew in my heart and soul that one day, I would be that guy.

## ⚔ LEARNING THE DRUMS AT THE STARWOOD ⚔

The Starwood was a famous rock club at Santa Monica and Crescent Heights Boulevard. Van Halen and Quiet Riot played there all the time, as well as lesser-known acts such as Y&T and the Quick. Our first time there, we just slipped in the door. Once inside, we checked out the place and walked right up to the VIP room, pulled the curtain back, and saw a band called London playing. I vividly remember seeing Nikki Sixx onstage; his hair was spiked high up in the air. He was dressed in black leather, and he was playing a black-and-white-striped bass. It was the coolest thing I ever fucking saw, right up there with the Orr sighting.

That experience was so amazing, so new to me, that I started going there every day at two or three in the afternoon. When the bands rolled in to set up, I'd help them move in their equipment. I just started hanging out and soon became a regular. Saul wasn't into it as much as me, and it became sort of my private thing. When the bands played, I would go up this stairway that led to the backstage area that the bands used.

In this area, between the back wall and where the drums were set up, was a small space about a foot wide. I could squeeze right in there because I was skinny. There was a little crack in the wall and from that vantage point I could look right down on the drummers. I'd study their every move, and that's how I started to learn the techniques of playing, from watching the pros. I was just a couple of feet away, and I could see everything. Sometimes my foot would fall asleep, or my back and neck would start to cramp up. I didn't care because this was the greatest possible opportunity for me. I honestly believed I was blessed to find this secret place where time seemed to stand still.

## ⚔ MEETING THE BANDS ⚔

I hung out and talked to everybody. I remember the Go-Go's playing there in late 1978. Belinda Carlisle had a shaved head and was just a happy, chubby-cheeked girl. She definitely got her act together in the following years. I met a lot of musicians, but the encounter I recall best is meeting Danny Bonaduce. I was tripping my balls off on acid, which was a new pastime for me that Saul wasn't particularly into, and I walked into the management office. He was sitting there with a bunch of coke on the table. I was at the point in my acid trip where there wasn't any barrier between what you thought and what you said, so I just blurted out: "Dude, you're Danny Partridge!"

"That's right," he replied, totally deadpan. I was so happy to see this person I watched on TV all the time, and I just smiled at him like a blithering idiot. There may have been a long awkward period of silence, but I don't remember. I know at one point I was back outside, probably leaving Danny to do his thing.

After hanging out there every day, the owners got to know me and I had free rein in the place. It was the seventies, and I couldn't help but feel that everybody was carefree, partying, and having fun. But it wasn't always like that. In fact, some bad, traumatic things happened to me during this time.

## ⤝ END OF INNOCENCE ⤞

The managers of the Starwood were these men in their midtwenties. All they wanted was to do drugs, fuck, and party. They were extremely smart, eccentric hippie white guys. They were all gay, and I was this cute blond-haired fuck-boy. They just loved me. I wasn't into that, but I was young and naive.

They would give me quaaludes and have their way with me. I just wanted to hang out, be with people, and enjoy life. But when you're young, doing your own thing roaming the streets, crazy shit happens. I ended up doing a lot of things that I didn't understand or really have any control over. In retrospect, a lot of things happened to me that probably messed with my head and hung over me for years, particularly when I found out that these young men later died of AIDS.

Just walking down Santa Monica to the Starwood or to Saul's

house, people would pull up beside me in their cars and ask me if I wanted to smoke a joint. I'd be like, "Hell, yeah!" The next thing you know you're completely baked and they're touching you all over and you don't know what the fuck's going on. All you know is that an orgasm feels good. Anybody can make you come, and in that state I didn't have the presence of mind to give a damn. I was used, abused, whatever. Let's get high. Let's party.

One time I was walking along Santa Monica Boulevard and ran into two clean-cut guys who must have been in their twenties. We started talking and they said they had some bitchin' weed back at their pad, so I went with them to smoke.

We arrived at this dumpy little apartment and there was another guy there, only he was in his forties, a completely scruffy-looking loser. Right away, I felt uneasy. Something wasn't right. This guy got up and locked the door behind me. "You want some grass, kid? Well, I want something too . . ."

The younger guys weren't friendly anymore. They slipped behind me while the loser walked up to me and ran his hand through my hair. I'll spare you the ugly details, but they hurt me pretty badly. Part of my mind just kind of shut down, and that day my reality became a bad dream. They didn't beat me up, but they did everything else and it was pretty devastating.

I was just fourteen at the time. I went home, stripped, and got in the shower. It was only then that I realized I was shaking pretty badly. After cleaning up I went out and got real high. Party, laugh, onward . . . and never tell a soul about it, until now.

## ⪦ ORGIES AND ORGASMS ⪧

It was a time in Hollywood where the overriding attitude about sex and drugs was to be free and out in the open. There was no panic over herpes or AIDS, no worries. Saul and I would hang out at Osco's Disco on La Cienega just across the street from the Beverly Center.

Of course, it wasn't normal for fourteen-year-old kids to hang out at a disco, but we had an older look to us. And even if we got carded, Saul, who was an expert artist, had taken our IDs and changed the date to make us of legal age. We never had a hard time getting alcohol or getting into the place. We went there just about every weekend of 1977 and 1978. There were like ten different theme rooms in the place. They were mostly sexual in nature, with settings like the baths of ancient Rome, open deserts with rolling dunes, a fully equipped dominatrix chamber, the wildest shit.

Upstairs and downstairs, everybody was doing coke and something called "rush," the popular drug fad at the time. It came in a bottle, and you were supposed to remove the cap and inhale the vapors from the liquid while you were having sex. By raising your heartbeat to insane levels, rush was supposed to intensify the orgasm experience. Throughout the night, we would go through all the rooms. It was an eye-popping experience. We never got bored, and if things started to lag, we'd just pop to another theme room. It didn't matter; they were all saturated with booze, drugs, and blaring disco.

It seemed like the more crazy the spectacles we witnessed, the more we hungered for wilder, more perverse thrills. Nothing could shock us anymore. Our nerves were deadened to the point that we stood there watching a three-on-one with the girl servicing every sick whim, only to be manhandled to the point that it was a borderline rape, and we'd be like, "Whatever. Next."

## ⇥ GAY SCENE ⇤

All the gay bars were along Santa Monica Boulevard, and most of the area's neighborhoods were predominantly gay too. I remember hearing Queen's song "Another One Bites the Dust" a dozen times every day. The Boulevard was definitely the primo gay hangout.

And the Starwood was the number one gathering spot for everyone. We goddamn lived there. We saw a lot of things that I wouldn't have seen back in Cleveland—guys getting sodomized in alleys or getting blow jobs from other guys in public bathrooms. Everything was out in the open and people were so into it. Saul and I were witnessing the raw, unbridled climax of a very narcissistic, very adventurous, experimental time.

Eventually, I had enough. It wasn't any one event, just the culmination of too many sick, beyond-the-limit nights. I needed to slow it down. I had weathered too many mornings waking up in someone's backyard with no recollection of how I got there or what led up to getting there. I did what any fucked-up Jewish boy would do once he's realized how empty and pointless the whole world has gotten. I moved back to my mom's to give home life another chance.

I called my mom and told her I was feeling homesick, that I wanted to come home. Things got very emotional. At first she hardly even recognized my voice, but when I heard hers it was like this wonderful oasis had come back into my life. She said very little, because I don't think she wanted me to hear her break down. She told me to come right home, and that was all she had to say.

## ⊰ MOM ADLER ⊱

I want to say this right now about my mother. I have never hated or loved a person as much as I've hated and loved my mom. I have never put another human being through the torture and abuse that I've heaped on my mom. I have never hurt or disappointed anyone on this earth as much as I've devastated Deanna Adler.

And yet I have never experienced anyone with a bigger heart or a more immense capacity to forgive than my mother. To this day we have our conflicts, and I continue to harbor a hatred for her that is all out of proportion with reason. And yet I've never depended on anyone or loved anyone so deeply. Deanna Adler has come through for me again and again and again.

My mom has said, "I love Steven, but I don't like him." And I believe I've given her no choice in that matter. I'm not a good son, and I'm not always a good brother, but that doesn't mean I can't give it another shot.

That day I called, my mom took a bigger chance on me than she ever had before. She sure didn't have to. She had peace in her home. She had a loving husband in Mel and a sweet joyful son in Jamie. She had a happy home and a stable, decent life. And yet she risked it all for me. Although she knew it would upset him and probably lead to a bitter argument, she talked to Mel about me. It had been several years since I had been kicked out, and I was hoping time had softened some of the harsher memories. Using what I can only imagine was a lot of pleading and love, she convinced my dad I deserved one more chance to be back in their home.

## ⊰ THE BIG DAY ⊱

Saul had found a pair of leather pants in a Dumpster by his grandmother's building. We cleaned them up, and he let me borrow them. I was the only kid wearing leathers at that time. I thought it looked very cool, especially with my long blond hair.

It was a look, however, that could not have been less popular with my parents. I tried to be optimistic about my return home, and so did they. Jamie was so happy I was back and for that first day, for that first meal together, I knew how much a family's love could mean.

Their demands were minimal and, in hindsight, pretty reasonable. They just wanted me to be a normal son. They asked that I be in by six o'clock for dinner and stay home evenings during the week. They asked me to keep my room in decent shape and clean up after myself in the kitchen and bathroom.

Almost immediately there were a few ominous signs. I was forty-five minutes late for the second dinner we had together. I blamed it on Saul. I told them I didn't have a watch and had asked Saul to let me know when it was five thirty, but he spaced and by the time I remembered what they said, it was already six thirty.

But hell, I was used to not coming home at all. When I walked in, it was very uncomfortable at the table. Mom had insisted they wait until I came home to eat and had tried to keep dinner warm in the oven, but the chicken ended up being all dried out. It tasted like sawdust and only served to increase the tension in the air.

Mom kept saying, "Why can't you come home on time? That's all we ask. Why can't you come home on time?" And Mel was all, "Stop hurting us. Stop hurting your mother. Look what you're doing to her. This has got to stop."

# BUSTED DRUMS, BUSTED FACE, BUSTED

I was fifteen now, and despite pissing off my parents by being late for dinner, I sincerely wanted things to work out at home. So when Mom told me she had enrolled me at Chatsworth High School after I settled in, I did my best to be a model student. It all came down to my ability to get along with people, whether they were in the dining room or the classroom.

During my hitch at Chatsworth, I met Dan Scheib. He was the grandson of Earl Scheib, the famous auto paint guy. Dan was a decent guy who played guitar but because he came from money, he had every other instrument in creation lying around his house. So I talked him into selling me his drum set. I even tried to jam with him when I went to check the set out. It was my very first time behind a real drum set and it didn't go too well.

Dan told me I could have the set for a hundred bucks. But when I showed up to pay, his father walked into the room and told me he wouldn't allow me to take the cymbals unless I gave him another $25. I looked at Dan, but he had his eyes glued to the floor.

I couldn't believe it. His old man was fucking loaded but it didn't matter. What a prick! The rich eat their young before parting with a dime, and old man Scheib just looked at me like I was some lowlife shit weasel, a flea market refugee to be bullied into giving him more than I had agreed to pay.

What could I do? I wanted a real drum set more than anything else in my life, so I anted up the extra dough, took everything home, and set it up in my bedroom. Giving Scheib the extra twenty-five turned out to be a pure waste of money because the cymbals cracked the first time I played them. They were such pieces of shit, they didn't even have a brand name on them.

As for the drums, I didn't know what the fuck I was doing. I was great on books and pillows, but now I had a drum set where you had to coordinate one foot pedal with the other foot pedal and both hands, and I was like, "Oh my God, what do I do now?" So, I visualized myself at the Starwood, putting myself in that same hallowed spot where I had seen all the drummers perform. I put on eight-track tapes of the Doobie Brothers and Boston, closed my eyes, and just listened. After a bit, I tried to play along smartly, because I didn't want to let down my fans.

Or I should say "fan," because Jamie was the only one there and he was my audience. In his eyes, I could do no wrong, and that must have helped, because I quickly got the hang of it. And just like your first car, you never forget your first instrument. I played the hell out of those drums, pounding on them with such ferocity that I'd break the whole set down to the ground during each practice, then pick everything up and reassemble it all. A couple weeks later, I jammed with Dan Scheib again, and you know what? His eyes popped. I could tell he thought that I had gotten a lot better.

## ⇥ BUSTED ⇤

My first week at Chatsworth High, I met a girl named Lisa. She was a pretty hot-looking senior with long brown hair, a big smile, slender legs, and cute little teacup tits. I had to ask her out.

She had a car and a job, which meant she was the perfect catch for a sophomore like myself. Of course Mom and Mel's rules still applied, so I would sneak out of my bedroom window after dinner and meet Lisa. My heart would be pounding so hard, giving me the best pure high. I miss that more than anything else—the great buzzes I'd get from the simplest things.

Before you know it, the booze and the drugs have muscled their way in to replace the natural buzz, then they make it impossible for you to enjoy any kind of joy at all, because you're all torn down, dead inside. You have no clue how you got so out of touch with the simple pleasures. If kids could just maintain a knack for capturing the natural highs in life, then the other crap would turn them off. Drugs would just bore them shitless when compared to a genuine high.

I'll never know how long I would have lasted at Chatsworth because I got blindsided only six weeks after I enrolled. Well, I guess that's my answer: six weeks. The only positive thing was that I happened to be there just long enough to get my driver's license out of driver's ed class. Back then, you could get your license when you were fifteen and a half.

I never saw it coming. I was in homeroom the day two police officers came into school asking for me. Some fucker had narced me out for smoking weed before class. Such a fun way to start off the day: being led away in handcuffs in front of my classmates.

You never forget your first kiss, your first car, or your first time in the slammer. I wasn't pissed—okay, maybe a little—but I was mostly stunned. The cops were going all Joe Friday on me, making it sound like I had just been caught with forty kilos of primo bud in my locker. After spending what felt like a whole day in a holding cell, I started to get worried. Maybe these guys actually were determined to make an example out of me. With my luck, it seemed possible that whatever the maximum penalty was for getting high before school, I was going to find out. Turned out I was only locked up at Devonshire police station for three or four hours before they called my parents, but it seemed like forever.

When Mom finally came to pick me up, the police captain told her that there would be no record of the arrest, that they were just trying to give me a good scare, which I guess was their policy for first-time offenders. I, of course, didn't know this, and when I got in the car, I went a little overboard with the "Please forgive me—it'll never happen again—I've been such an idiot" routine. But it didn't really make much difference because Mel had already issued his edict before Mom came to get me.

Mel said no way I was welcome back in the house, so I was taken to a foster home in Pasadena. There were all these crazy kids running around, screaming at the top of their lungs. I never felt sorry for myself, it was just "This sucks. I'm out of here." So I called Lisa and said, "Woman, you gotta come rescue me." Within an hour she was there to pick me up right outside the home. The lady running the place was freaking out, all yelling after me: "You're not allowed to leave. Come back here. I'm calling the police!"

## ⚔ MY DESTINY DEFINED ⚔

We were just like Bonnie and Clyde. We laughed and tore off down that dirt road. I can still remember the unbridled freedom I felt as we sped away. That night I realized that home for me would always be where I was loved for what I was, not what I could be molded into. I told myself that home was an illusion anyway. We all die alone, and the people who learn to be their own good company are the ones who have the best shot at being happy.

Lisa was so very cool. She was always up for anything, and I loved hanging with her. We would drive to Malibu Beach and park up by the cliffs. We'd party in the car, push the seats down, and slide them way forward. We'd screw in the backseat and then pass out.

That summer Lisa took me to the U.S. Festival in San Bernardino. We went on Sunday, the heavy metal day. Opening the show was Mötley Crüe, who were well on their way to becoming superstars. I remember studying every move Tommy Lee made that day—the way he counted off the next song; the way he twirled his sticks;

the way he used those sticks to point at the audience, then at the other members of the band. Those drumsticks became extensions of his hands.

Most important, I studied the way Tommy smiled that day. He was always beaming. I got chills. That was gonna be me. I couldn't stand musicians who looked so serious up on the stage, like they were constipated. If you want to look that way, go get a fucking root canal. Playing music is a pleasure and a privilege. Tommy showed me the way, and that was to be my way. Thanks for being an incredibly positive inspiration, Tommy.

## ⊰ NO REPRIEVE ⊱

They'd only let me back in school if I agreed to see a counselor. He wanted to know about the recurring trouble at home and why I was skipping school. Even though Mel and Mom had relented and let me back in the house after the foster home fiasco and my weekend furlough with Lisa, it just wasn't working out. I was just impossible and told the counselor so. I said, "Life at home sucks. I can't stand it and I can't stay there. If you can pay me three fifty an hour, I'll get my own place and guarantee a C average."

He said, "I can't do that." So I told him I couldn't stay in school; I needed to work.

I had been home less than three months before I was kicked out and all my shit was on the street again. I should have asked Mel to install a revolving door on the front of the house. At least there were no extended tirades this time; it was just "Hit the road, loser."

I couldn't take the look on Mom's face. Now, it's not like I hadn't seen it before, I just didn't want to see it again. It meant failure, total failure, my failure. My repeated ability to screw everything up pummeled my parents to the point that now they just went through the motions without any feelings.

Mel and Mom had most of my stuff out on the front lawn before I got home. The worst was saying good-bye to Jamie again, his teary eyes, wet cheeks, trying to smile through it all. I don't even want to think about it. He was standing in the yard and I knew if I went over to hug him, I was going to lose it. It was déjà vu all over again, loading up Stormin' Norman's troop carrier to take me back to Big Lilly's, all four feet six inches of her.

Maybe it was because I had blown my last chance, burned all my bridges, and failed my family for the final time, but this banishment was the toughest to take (even though I pretended it was no big deal). For the first time I felt a distinct chill seep into my heart, an emotional vacuum that told me I would never be welcome back. No amount of time would heal this rift; forgiveness was no longer an option.

## ⊰ AS IS ⊱

So there I was, back at my grandma's. It was time to suck up to her big-time, because I realized there was no longer any safety net; she was the only thing between me and the streets. But when I looked at her serene smile, I realized it was all going to be fine. I could do no wrong in Big Lilly's eyes and besides, she was in good spirits during this time. I guess it was because she hadn't seen me in a while and, with her waning powers of recollection, had forgotten what a horror show I could be.

I decided I was going to make the most of her good mood. Within a month, by sucking up and doing small odd jobs around the house, I had bribed enough money out of her to buy my first car. It was a Chevy Corvair. It cost me $200, "as is." Boy did I ever learn the meaning of "as is" in the classified ads the hard way. I bought it from some greaser dude in North Hollywood. He really did look like he combed his hair with axle grease.

I drove it out of the driveway and onto the road. I was smiling, planning to wheel by Saul and impress him with the new party wheels. I was already scamming on how to get enough money to stick in a decent sound system.

I made it about a block and a half down from the greaser's place and the car caught on fire. I was so pissed, I got out and walked away. I just left it there burning in the street. It was my own eternal flame, a torch tribute to my stupidity. But I figured the people in the neighborhood knew who it belonged to, so let that asshole take care of it. That was one quick $200 wasted.

Out of the eternal goodness of her heart, Big Lilly took pity on me and gave me her old car, a blue '75 Gremlin. The Gremlin was put out by American Motors and is one of the ugliest cars ever made. I used to think that American Motors only hired blind people for their design department. Otherwise, how else could you explain a car company that was responsible for freak shows like the Marlin, the Gremlin, the Matador, and that fishbowl on wheels, the Pacer? The Pacer was 90 percent glass. It was impossible to toke up in a Pacer. There was no place to duck behind if the police cruised by you. Hell, they could spot your roach clip in the ashtray.

Now, I have no idea why, but Big Lilly's simple act of generosity got my whole family in an uproar. Everyone was against her giving the car to me. But what they really resented was that I was Big Lilly's favorite. One day Mom's brother, Uncle Artie, came over and ripped all the wiring out of the Gremlin because he was so pissed that she gave me the car. He must have done it at night, because when I went to start it up the next day, it was completely gutted.

Word soon circulated as to who the culprit (hero) was. And later, when I confronted him, he freely admitted it. This is just typical of the whacked-out relationships that ran through my entire family. So what the hell chance do I have to be normal when half my family was slaughtered by Nazis and the other half can't bear to let anyone show the love?

But in the end it was totally cool because my grandma put all new wiring in it, and it ran even better than it did before. And that sucker could run. It had a stick shift and a six-cylinder engine that flew. I think it gave Grandma some pleasure to show that she couldn't be bested.

Later, Saul spray-painted "Road Crew" on the side of the car. He used a cartoon he drew of a girl clenching a flower in her teeth, señorita style, while her boobs hung freely. Done in white paint, it looked pretty fucking cool. Later he got that same design tattooed on his arm. He was an amazing artist. We dubbed that Gremlin the Road Crew-zer.

## ⤞ GETTING MY DRUMS ⤝

Now that I had wheels, all I needed was a professional set of drums and the world would be mine. I found a gorgeous set of drums I really liked at Granada Hills Music Store. But they were going to cost $1,100. Before I could pick up the drum set, I had to pay for it. So I would go back there every week and put a little money down until it was paid off.

The store was actually a house on Fairfax and Santa Monica. Saul took guitar lessons every day there, and I would go in the drum room and mess around with the drums, getting a feel for them. The salesperson at the store showed me the intro to Ozzy's "Over the Mountain." I thought, "If I can learn to play that, I can play anything."

One time we were at the Fairfax Music Store, where Saul was checking out the red BC Rich Warlock that he later ended up buying. Mike "Flea" Balzary walked in and he picked up a bass off the wall and just started slapping it like a pro, unbelievable. We were like, "Dude! When did you start playing the bass?" and he smiled, saying, "Two weeks ago." That was crazy shit. We were very impressed. Before they formed a band, we used to see the Chili Peppers, Flea and Anthony, hanging out around town all the time.

I wanted those drums so badly, I started taking on any kind of odd job. Anything. Soon I was juggling four different gigs. In the morning, I'd make dough and prepare the sauce at a pizza place. At night, I bussed tables at a country-western bar and grill, and I worked at a 7-Eleven from ten at night to six in the morning every weekend. There was a burrito place next to the 7-Eleven and every day Saul and I would get a burrito and a taco there. I also worked at a self-serve gas station on that same block from ten P.M. to six A.M.

weekdays. I'd get so tired, I felt like I was hallucinating. But I would just keep going.

I kept hearing those drums and how great they were going to sound. Wait a minute, I mean how great *I* was going to sound. Finally, I saved up enough to get the drums, but I still needed another seven hundred to get the cymbals I wanted, and it wasn't easy, because I tended to get fired from every job I ever had.

### ⇥ NEW DRUMS ⇤

So my second set of drums was a Tama Rockstar. It had two twenty-six-inch bass drums; fifteen-, sixteen-, and eighteen-inch toms; one floor tom; and a wood snare drum. Now, if I could just get those damn cymbals. When I sat down, I looked over my drum set like it was my personal vehicle to the top. I knew I was going to be successful, never doubting that I would make it. I figured if Van Halen could do it, well . . . they eat, shit, drink, fuck, and jack off just like I do. So I guessed I could do it too.

Rock is like the blues in that you got to pay your dues. I reasoned that if I had one job, I'd be paying my dues; if I had two, even more. With four jobs, I figured, "God, I've really paid my dues in spades and that means I'm gonna make it one way or another."

Sometimes, Saul and I worked together. When we were sixteen, we were hired at Business Card Clocks. They'd take a picture of a business card, blow it up, put it in a frame, and put a clock on it. I'd stain the wood and make the frames, and Saul would shellac the business cards after they were finished. Fact is, we were just hired as extras for the added orders that came in for Christmas presents. Nothing says "Merry Christmas" like a business card clock!

Saul also held down multiple jobs, working at a movie theater on Fairfax, a magazine stand at Fairfax and Santa Monica, and a music store. Saul never stayed with one job very long, but the difference between us was that Saul, and not his boss, was the one who decided to move on.

When I finally got the cymbals, I had an entire professional drum set that inspired me to practice until I was pretty damn decent. Every second I wasn't working I was practicing. Hours and hours of exercises mixed with mimicking the drum parts from every great song I loved—Ozzy to Aerosmith, Stones to the Crüe. I set the bar pretty high for myself; I wanted it all to sound perfect. Finally I built up the confidence to call Saul and say, "I've got it together." We arranged a rendezvous, and I packed the drums into the Gremlin and drove to La Cienega Park just north of Olympic. It felt like the perfect spot, playing in the big wide-open in an area the size of four football fields. I had everything set up before Saul arrived around eight o'clock. With Saul and some of my other friends there, it almost felt like an audition. Hell, it was an audition, for my friends, for Saul, and for the gods of rock. They were all there that day.

I was never so happy. I had finally gotten my shit together, and I set it up for Saul to check me out and let him take it from there. I played a few beats, executed a few licks, then dove in, really giving it my all. After about a twenty-five-minute salvo, Saul was impressed enough to say, "Cool." From that moment on, we really locked on to "the Dream."

I started to reach out to successful musicians because I wanted to surround myself with performers who were not only inspirational but possessed talent and drive. I met Robbin Crosby, rhythm guitarist for Ratt, at the Rainbow. After an eight-year battle with AIDS brought on by drug abuse, he passed away in 2002. Great guy. RIP, Robbin.

Robbin was huge, six and a half feet tall, and good-looking. He took me under his wing and decided one night to take me over to Carlos Cavazo's house. Carlos was the guitarist for Quiet Riot. QR was amazing. They had the largest-selling heavy metal debut album of all time, until my band took that honor a few years later.

Carlos lived in Laurel Canyon, right behind the elementary school there. Ratt's vocalist and drummer, Stephen Pearcy and Bobby Blotzer, were also hanging out there. It was a hell of a night.

Seeing all the platinum records on the walls was awesome, and I never doubted that I would soon have my own. We just drank and partied all night. There were always freshly cut lines on this shiny, slick wooden table in the living room. I was freaking the fuck out. This was the famous debauched rock 'n' roll lifestyle, and it was awesome.

After a night of partying I totally lost track of the time. I asked the guys, and Carlos laughed and pointed at a clock. It was six in the morning. "Shit. I have to get to work!" At the time, I was working for a poster shop where I would spray the glue on the backboard and they would mount the poster onto it.

On my way to work I walked through a garage for an apartment building. I was so tired. I needed to rest for a moment. I just went down the rows, and after two or three cars, I found one that was open. I got in the backseat and fell fast asleep. Nice, ahhh . . . then . . . "What the hell are you doing!?" The owner of the car was pounding on his window and shouting at me. He was going to work in a suit and tie. I was shocked awake and asked the guy what time it was. And again he shouted: "What the hell are you doing!" Then he looked at his watch. "Seven fifty."

If I hadn't been so tired, I would have laughed; it was a pretty funny scene, Mr. Hangover meets Mr. Suit. "Shit, I'm late for work." I got out of the car in front of the startled businessman and ran to my job, only to find that it wasn't waiting for me. I tried to explain what had happened, that I was exhausted and very sorry to be so late, but they looked at me like they didn't know me. I was fired, gone.

## ⇥ OH WELL ⇤

While living at Big Lilly's, I rarely saw my mom. Passover dinner was usually at my aunt Greta's house. She was my mom's sister and provided one of my few chances to be with the whole family each year. Visits with my relatives seldom lasted though. Some uncle would make a remark about my hair or my being fired recently, and

I'd answer back with some over-the-top rude comment. They'd gang up on me, and soon I'd find myself being asked to leave. My relationship with my family continued to flounder, because I just wouldn't shut up and take their abuse. I think that's why I so enthusiastically embraced GNR as my family; they accepted me just the way I was.

If GNR was to be my family, then Saul was my brother. We were really getting into our music more and more, rehearsing all the time. We had been hanging out with a good friend of Saul's, Matt Cassel. He lived on Sunset up by Carny's Diner, the hot dog joint that looks like a train. His dad's house was just above there. On his property stood this enormous tree, which grew out of the side of the hill facing Sunset. It had a makeshift swing, two ropes attached to a wooden seat. I would get stoned and fly right out, way above the Strip. It was awesome fun.

## ⊰ THE BIRTH OF "SLASH" ⊱

Matt's father is a professional actor named Seymour Cassel. He's been in some great movies, like *Colors, Rushmore,* and *The Royal Tenenbaums.* Saul and Melissa spent a lot of time hanging out with these guys. When I would go over there, I noticed how Seymour would always call Saul "Slash." It was just his personal nickname for Saul and for some reason, that really stuck with me. I couldn't forget it because it just seemed to fit Saul so well. The name "Slash" must have resonated with a lot of people, including the man himself. After a while, Saul made it known that he had taken a keen liking to his new name and the rest is history. He told everyone, "Call me Slash." We were like, "Slash? . . . Done." From that day onward, that's how I, and soon the world, would know him.

By the early eighties, I had already been living a rock 'n' roll lifestyle for several years. I was running wild in the streets of Hollywood, partying with rock stars, fucking all kinds of hot, crazy girls. I was in and out of dozens of odd jobs during this time and spent all my free time either practicing or going to as many concerts as I could sneak into.

By this point I had met a shitload of people. I just kept networking, meeting the characters who were living the life I wanted. I always had a mind to see if they could help me with my music. It wasn't like I was looking to use them, but if they knew a club owner, or could get me a deal in a studio, or knew a pawnshop where I could get a break on a cymbal or something, I put them on my short list.

Slowly but surely, I was moving up. Armed with my Tama kit, a positive attitude, and a "new do," now cropped and spiked, doors were opening left and right for me. It was because of the presence I could bring into a room. I acted and looked the part, and I could back it up with the best drumming in town.

## ⊰ WORKING ON THE DRUMS ⊱

In December '82, I found a room to rent in the home of my friend Brad Server. He was one of those surfer dudes who love Southern California, the epitome of the Jeff Spicoli character from the movie *Fast Times at Ridgemont High*. He lived with his mother down the street from my mom. She owned a big three-bedroom home. I would stay there a lot in the spare bedroom for only $125 a month. Brad's mother was the daughter of Curly, my favorite of the Three Stooges. It was just the two of them there and I was allowed to set up my drums and jam. During the day, Brad would go to school, his mother would go to work, and I had the house to myself.

So I would just practice all the time. I remember I would play to Journey's *Escape*. I loved that record. They had the greatest drum sound and Steve Smith was damn good. I had Ozzy's "Over the Mountain" down by then too. This was the time that I made some of my greatest strides on drums. I stayed there for a few months and I appreciate Brad and his mom's hospitality to this day. A rocker never forgets the people who help him out when he's a nobody.

In January '83, I took Lisa with me to the Rainbow. The Rainbow was to become our second home. It did not discriminate between big hair, short hair, rich, poor, famous, infamous, rock stars, roadies, drug dealers, record execs, wannabes, and hangers-on. The "Bow" welcomed us all.

Lisa was the closest thing I had to a steady girlfriend, but of course, I was fucking around a lot too. I had been going to the Rainbow for years, but never once had I brought a girl there. The Rainbow was a place to get girls, not bring girls. Lisa and I had the small booth in the back right corner. At one point I got up to go to the bathroom and I got stopped at every table. Chicks I knew and didn't know all had me sit with them.

I was having a great time, just swinging from one table to the next. I literally made out with a different girl at every booth. So I didn't get back to Lisa for a while. When I finally returned, Lisa was freaking out on me: "Where the fuck did you go? I've been sitting here for an hour." Like I explained, I'd never once brought a girl to the Rainbow with me and now I realized why. It really cramped my style.

I'm thinking, "What? I was with some chicks." I didn't understand or even comprehend the idea of being in a serious relationship. She was upset and wanted to go right away. As we were leaving, she's yelling at the top of her lungs. We're walking out, and she's screaming that I'm an asshole. I was pretty drunk, and I suddenly became very aggravated. I turned around and yelled, "Shut the fuck up."

We were right at the main entrance by the cash register when all of a sudden some big-ass guy grabs me, turns me back around, and punches me right in the face. I don't remember *anything* after that. But when I came to again, I had evidently gotten into my car and driven to Mom's house.

Canoga Park is over twenty miles from the Rainbow and I had absolutely no recollection of the drive. The next morning, when I woke up, I was still in my car, which was parked in front of Mom's place. As my mom was leaving for work that morning, she was shocked at the sight of me and started banging on the windows of the car, where I had passed out in the front seat.

There was blood all over the seat, my face, and my clothes. It was the first time I had had my nose broken and the feeling was terrible. I couldn't breathe and at first I had difficulty focusing my eyes on anything close up. I looked like Marcia Brady in the *Brady Bunch* episode where she got hit square in the nose by a football.

My mother knew just what to do. She muttered something about needing to fix it right away, and I was damned lucky I was under eighteen years old, because her health insurance policy still covered me. She rushed me to the hospital and got the doctors to look at it right away. Within twenty-four hours they had operated, and somewhere there's a photo of me smiling from my hospital bed with my nose all taped up. Mom should have gotten a twofer rate, because it wouldn't be long before I'd whack my schnoz again and need another nose job. That was just one part of my body that would be mangled and fixed repeatedly over the next twenty years. I'd like to thank all of the doctors, nurses, family, and friends who have carried me off the battlefield and treated me a lot better than I ever treated myself. It's a miracle that I'm alive, but in my early teens I believed I was indestructible and probably didn't even notice the self-abuse until my first overdose.

# 6

# THE BIRTH *of* GUNS N' ROSES

Gardner Park used to be a big empty storage warehouse where they held auto shows in the 1930s. I found a place to set up my drums there and would play for hours inside, where the acoustics produced a big John Bonham sound and the echo effect was like the intro to "Misty Mountain Hop," massive, very awesome. The place had been deserted for years and grass had squeezed up through cracks in sections of the cement flooring. I could play, but the set would wobble; it was so unstable my cymbals would rock back and forth. I had two giant Asian gongs, more than three feet in diameter, an idea I ripped from Carmine Appice, who had the coolest setup when he played with Beck, Bogert, and Appice.

While I was practicing one afternoon, one of the gongs must have shifted because of the uneven flooring and came flying down. I looked up at exactly the worst moment and took the full brunt of thirty pounds of metal in the face. It knocked me clear off my stool and onto the hard floor. I took off my T-shirt and wrapped it around my head over my nose, then found myself back at the hospital getting my schnoz rebuilt again.

During this time I was working at the O'Neal Motorcycle Shop warehouse, where I printed the O'Neal logo and numbers on T-shirts for bike riders. The floor manager there was a guy named Mark Marshal, a cool guy and a great guitar player who looked like a musketeer with his goatee, long black hair, and long thin pointy nose. Eventually, I got fired from O'Neal for always being late. But by then, Mark and I had become good friends, and we agreed to form a band. So with a bass player, a guy I think was of Russian descent, we all got an apartment together. But even though we really wanted to start a band, our schedules didn't allow us much time to get together.

I found another job, but we were all basically broke, living on nothing. I remember eating melted butter over steamed rice every day for a week. We didn't even have soy sauce. No one's going to go out and buy a bottle of soy sauce when you're saving up for a Def Leppard concert.

In April '83, Mark and I went to see Def Leppard at the L.A. Forum. Even though we were skimping and saving for weeks, we still didn't have enough money for tickets, so we just went to the back entrance where the trucks went in. I have this great memory of hearing the song "Photograph" being performed. After the show, we just started helping the roadies load shit in the trucks, and the band came out and stopped by the first truck. They were standing right next to me.

I thought they were going to kick us out, so I figured, "Now or never, I gotta do this." I said hi to Rick Allen and shook his hand, which at the time was definitely the biggest rush of my life. I didn't get to see the show, but I got to meet Rick. I told him the story when we had dinner together years later.

Mark and I also saw a lot of cool bands at Chuck Landis's Country Club in Reseda, which was right down the street from our apartment. We saw the Christian metal band Stryper there a couple of times. They were so hot, really had their shit together, and drew huge crowds. I borrowed a couple of moves from their drummer,

Robert Sweet, who had this huge drum set and was set up sideways so you could see him playing.

"The Visual Timekeeper," he called himself. He was the coolest-looking motherfucker up there. Their music was so loud and clear that they sounded like a studio recording; it was that perfect. Oz Fox was incredible, playing guitar and singing backups. They looked huge, like Kiss, larger than life, with matching yellow and black outfits. There must have been some serious cash put into their show. I saw them three times and I loved them. Hey, I was a fan.

We saw Joe Perry at the Country Club too, during the short time that he was not in Aerosmith. Coincidentally I remember, years later, Axl told me that the first concert he saw was Aerosmith during the same year when Jimmy Crespo was playing guitar, and Axl thought his solo was one of the best he ever heard. The band Rose Tattoo opened up the show, turning Axl on to them and inspiring him later to have our band perform the Rose Tattoo classic "Nice Boys."

## ⊰ DIGGING DOWN DEEP ⊱

After a few months of living in that little hovel, I decided to move back in with Grandma. It was great going to all those concerts, but the band I had intended to put together with Mark and the other guy wasn't really going anywhere. So there's Slash and me, both back at our grandmothers' places now, working at any odd job we could grab. This was another low point for me emotionally, but it spawned a kind of simmering desperation deep inside, a fierce, burning desire to get it the fuck together. Slash and I reunited, drifting rogue musicians in search of the ultimate band. There was a dire immediacy in our playing now, and for the first time, we practiced together regularly. We started to gel more, and it was in those lean days that our sound and style really started to come together musically. We jammed nonstop, loud and proud. We would make the gods hear us; we would make the gods sit up and take notice.

One evening we were walking in front of the Roxy when I spotted a flyer on the ground and picked it up. There are a million band flyers floating around Sunset at any one time, but this one caught my attention. It was for a band I hadn't heard of called Rose, and they had a gig at Gazzarri's the following Tuesday. The flyer featured a picture of two guys standing together. They definitely had the look, the right image that was so important to the local rock scene of the time.

Although I had never heard of them before, I immediately felt in my gut that they had superstar potential. I showed the flyer to Slash and right then I said, "I swear, if we get these guys and a cool bass player, we will have a kick-ass band!" Slash nodded slowly, I think initially just to blow it off, but then he smiled. At that moment I believe he knew I might have been onto something.

The next Tuesday we went to see Rose perform. We arrived at about six o'clock. There were a lot of bands playing, so there was anywhere between fifty and seventy-five of each band's faithful listening during each set. The stage was sectioned off so there could be three bands' gear onstage at any one time. The guys in Rose were on the stage-right end.

It was a long event, band after band, like twelve of them. Rose got to play only three songs. I learned that the guys featured on the flyer who caught my attention were vocalist Axl Rose and guitarist Izzy Stradlin, two childhood friends from Indiana. I thought they looked cool and that even their names were cool. They had a guy named Rob Gardner on drums with them, but I wasn't that impressed with him.

The bass player's name was DJ. I believe he helped write "Move to the City," one of the songs they performed. He was skinny and had long black hair, good-looking rock 'n' roll kind of guy. But he was only in the band for a couple of months.

Just a few days later, I met Izzy Stradlin through my friend Lizzy Gray. They lived in the same building. Izzy and Lizzy had played together in the band London for a short time. Of course London was already notorious for launching the career of Nikki Sixx.

Izzy looked like a young Ron Wood, with that gaunt, angular cut to his face, perfectly framed by straight black hair that hugged his jawline, making his face look even more thin and elongated. He was into heroin, just like Ron Wood and Keith Richards, his heroes in the Rolling Stones (Woody had taken over for Mick Taylor by the time the Stones recorded *Black and Blue* in 1976). He had thick-soled platform shoes and always wore black pants with some sort of super-tight shirt. He looked more like his shadow than himself and to me was the personification of cool. Izzy and I hit it off right from the start. We each saw something in the other; perhaps it was just the way we talked about music. Izzy was the consummate rhythm guitarist. I loved the solid power chords he built into Rose's songs.

Izzy's apartment was below Sunset on Palm Avenue near Tower Records. It was a square little studio with a small kitchen and a tiny bathroom. We were hanging out there for the first time when I asked him about getting together to play. He was fine with the idea and he gave me their demo tape to listen to right on the spot. The cover featured the same picture from their flyer, and the cassette contained three songs: "Shadow of Your Love," "Move to the City," and "Reckless Life." I didn't get to keep the tape because Izzy only had two of them.

Since Rose had just gotten rid of, or was thinking of firing, Rob Gardner, we made a plan to jam together so I could learn the songs. Later that evening, I split and headed over to my friend Sue's; she happened to live right across the street. Sue was the sweetest girl, very accommodating, and her pad became a popular party pit stop.

## ⇥ MEETING AXL ⇤

Once, as I was leaving Izzy's place, I bumped into this skinny, pale-as-a-ghost rock 'n' roll dude with long orange hair. He was wearing a light blue, unbuttoned dress shirt, completely exposing his ribs, which were sticking out like a starving dog's. We met in the middle of the street, and I didn't recognize him at first, but when we said hey to each other, I realized who he was. "Dude, you're that kick-ass singer I saw play Gazzarri's. I thought you were great."

He smiled and said, "Thanks." He seemed very humble and gracious. That was the first time I ever spoke to Axl Rose. Apparently he was a regular at Sue's place and hung out a lot with Izzy.

Rose was rehearsing across the street from Hollywood High School on Selma Avenue at Selma Studios, which offered the cheapest studios, like $5 an hour. Now, that's a great deal, but it was one sorry shit hole. The building was ancient, with broken doorknobs, jammed windows, reeking bathrooms, creaky floors, and cracked walls, but you get what you pay for. I went to check out the situation without Slash because at the time, I didn't think they wanted another guitar player. It was just Izzy, DJ, and me. The first song I played with them was a number Izzy wrote called "Shadow of Your Love."

Axl sauntered in while we were running through it. Without missing a beat he grabbed the mike in the middle of the song and just started running up and down the walls, screaming and wailing like someone had lit his pants on fire. I had never heard such a sound in my life. It was like some otherworldly banshee cry. I was stoked. I remember my eyes bugged a bit and my pulse shot up; I was thinking this dude was so insane, so original.

I thought the session went off really well, and afterward we were sitting around talking and I said, "You guys gotta meet this guy Slash. He's fucking great!" And that's how it all started to come together, like we were each waiting in the wings of the cosmos, destined to discover one another, like it was meant to be.

I went back to the apartment and I told Slash, "Dude, these guys are great, they're totally original, very cool, and they want to meet you." Slash kind of made a noncommittal grunt, which you got to understand in Slash-speak is a very positive response. The next night, I brought him over to Izzy's just to hang out and see if we would all get along. We were drinking beers, shooting the shit, talking more about bands we loved. Axl was more talkative than the last time we hung out. He was saying how he was really into Dan McCafferty, the lead singer from the band Nazareth. I was familiar with their albums *Razamanaz* and *Hair of the Dog*, which had some kick-ass songs on it. If you listen to McCafferty's wailing, you can see why Axl was into him. He could really tear through a song and put a very distinctive brand on it, like his voice was another instrument. All the truly great vocalists do that; they make their voices a unique and indispensable part of the band's total sound. How the fuck do you replace Robert Plant, Freddie Mercury, Steve Tyler? Or, as was later to be discovered, Axl?

Izzy was playing his guitar and he let Slash check it out. Slash just kind of looked at the neck for a second and then tore off some cool lead stuff, but nothing too flashy, just brilliantly, undeniably Slash. Izzy and Axl were impressed enough to tell Slash to go and get his guitar. This was worth looking into and we could all feel the temperature in the room go up. Slash did just that, and soon enough we were jamming for another couple of hours.

That night, we all just happened to walk into one another's lives, with no idea what lay ahead. I wish I could say that it was like lightning struck, but the truth is that it was just a random get-together to see what could be germinating. Axl was the least vocal of all of us that evening, but when Izzy suggested we get together again, Axl's body language definitely indicated he approved.

The next day we went back to the studio. Izzy was now totally into the idea of another guitarist because he wasn't much of a lead player, nor did he want to be. Izzy could be freed up more by playing with someone like Slash. Like I said, Izzy's a very rhythmic, chord-loving artist and Slash is naturally solo oriented, so they connected in a real beneficial, complementary way, strengthening their styles by allowing each to focus on what he loved.

We pretty much just played those same three songs: "Shadow of Your Love," "Move to the City," and "Reckless Life." Then we jammed on some Aerosmith and Stones stuff too. We hung out for a couple days, and whenever we got some money, we went in and jammed again in the studio.

Rose, or Hollywood Rose as they were called on most flyers, had two dates they had already committed to, so Slash and I joined Izzy and Axl as the "New" Hollywood Rose without much fanfare. The first was at Madame Wong's East in L.A. Then we played at the

Troubadour on July 10, 1984. This turned out to be the first time my family saw me perform.

## ⊰ OUR FIRST SHOWS TOGETHER ⊱

My mom, dad, and brother Jamie showed up for the big night. It was great for me to show them that I was doing something other than being a fuckup. Seeing the looks on their faces was an incredible experience for me. They told me they loved it and were excited for me, although Mom said she was deaf for three days afterward. And poor Mel, he wanted to leave after the first song but rallied and hung out for the whole set. As for Jamie, he told me it was one of the greatest nights of his life.

We played the Troubadour again on August 29, 1984. We also played locally at a couple of parties, one of them actually a fucking kegger, which was in a tiny little apartment off of Highland. Everybody was cramped into a room that was so small that Axl had to sing sitting on my bass drum. We also played an after-hours party at Shamrock Studios, which was located at Santa Monica and Western. They would have the most insane parties there. The late El Duce from the punk band the Mentors would hang out sometimes. Slash and I would trip out on him. He was insanely scary.

I remember seeing a Mentors concert where a chick was giving him a blow job onstage. And at parties he never failed to surprise us. El Duce was *the craziest*. Once, he drank a whole bottle of Jack Daniel's right in front of us. He ripped the cap off a fresh bottle and downed it. He was chugging it like it was spring water. There were a few people around him with us, and we were like, "No way. He's going down. He's gonna fall flat on his face," but he did not go down. Instead, after he drained it, he shattered the bottle over his head. I've never ever seen a crazier motherfucker, but the booze finally took its toll when he was fatally hit by a train while drunk. Some believe he was shoved onto the tracks, but we'll never know.

For a brief period, Izzy left for what turned out to be a short stint in another band, L.A. Guns, and Hollywood Rose became me, Slash, Axl, a guitar player named Jeff, and another bass player, some French guy. Slash drew a cartoon of the new lineup for one of our flyers and it's featured in the *Live Era* record. There was no drama when Izzy left, because we had hit a lull and were spending more time partying than playing. The development of the band worked out very slowly, off and on over months and months.

More than anything else, we were just good friends cruising the Strip to meet chicks. The conversation was just like that of any other group of guys hanging out. Where were the girls, the parties, the best places to score bud? Who had any money, who was holding (we were always holding and never admitting it), and what were we doing later? There would be some talk of music, but most of that dialogue took place only when we were actually jamming.

We were out a lot. We were always at someone's place or at the studio. But most important, whether we were jacked in or not, and whether we had planned it or not, we were always *around one another*. When we weren't together it was because we were with our girls. Slash started dating this beautiful dark-haired babe. And I met this chick named Loretta who lived across the street from the Whisky. Our relationship was consummated immediately.

Loretta and I had been together for about a week when one evening she had a surprise for me. She was driving up in the Hollywood Hills in her '76 Dodge Pinto. "I know a fun dude who lives up here."

I asked, "Who?"

"Bob Welch."

"Bob Welch!" I excitedly shouted back. Welch was a member of Fleetwood Mac until 1974 and had a great solo career after *Heroes Are Hard to Find*, his last album with Mac. Two of his biggest hits were "Sentimental Lady" and "Ebony Eyes." "Sentimental Lady" was actually a song he first recorded with Fleetwood Mac on their *Bare Trees* album. Both versions are gorgeous love songs and needless to say, I was stoked to meet this guy.

Poor Loretta had no idea that Bob had OD'd on heroin a few days before and was presently in the hospital. His friend Ted was staying at his house, and he invited us in. Bob had a very plush, stylish home. We were hanging out in the living room when Ted lifted a glass pipe off the table. It looked like a clear lightbulb with two stems protruding from it. He handed it to me, and I held it in my hand and looked in the bowl of the pipe, and there was a little white chunk. I just figured, "Oh, coke." I lit it up and inhaled with no fear. It felt *amazing*, complete and utter euphoria. I thought it was the highest-quality blow I'd ever sampled. But it wasn't. High-quality coke would have been a tender blessing compared to this curse. Had that pipe been a loaded revolver, it couldn't have done more harm.

I had inhaled crack and exhaled my soon-to-be shattered soul. It was the first time I smoked the shit. As I sat there, an incredibly powerful urge came over me. I had never experienced such a dire need to get high again. Right away. *Now.* And this was only about ten seconds after that first incredible high. All I knew, all I cared about, was that I wanted the feeling to last longer. So I continued to hit the pipe. I didn't know it then, but at that very moment I had tasted the beginning of the end.

## ⊰ LENNON HATED BLOW ⊱

John Lennon once said that cocaine was a dumb drug because the only reason you do it is to do it again. Coke makes you feel like a new man. Unfortunately the first thing that new man wants is more blow. Lennon was able to walk away from cocaine after his famous "lost weekend" with Harry Nilsson during 1974 in L.A. He came back to New York City, begged for Yoko's forgiveness, and found happiness at the Dakota. John and Yoko's love was stronger than coke. Yoko bore Sean the next year and John's new high was being a "house husband," raising his son for the next five years. That's a perfect example of finding a natural high to supplant drugs. And to me, that's one of the most beautiful power-of-love stories in the history of mankind. I wish Lennon's example and willpower could have inspired me to quit blow right then and there. But I was weak and hungry for the elusive high.

Eventually, Loretta said she wanted to leave. I told her, "I'm not going nowhere." I was smoking that shit and repeatedly inhaling it with a suicidal urgency to maintain the high. We went up to Ted's bedroom, where he had a bottle of vintage Scotch. He broke the seal and poured each of us a glass. Loretta continued to pester me to leave. Finally I said, "Look, I'm having fun. If you wanna go, go." She split and I never saw her again. She knew it was hopeless. She knew *I* was hopeless.

Meanwhile, I'm figuring, "This is great, I'm not spending a nickel of my own money." Ted was hooking me up just to hang out. And misery loves company. Two miserable drug addicts stoking each other's habits. Within a couple of weeks I had set my drums up on some carpeting in Welch's garage and moved in.

## ⊰ INTRODUCING MR. BROWNSTONE ⊱

Right after I set up my drums, Ted took me into Bob's private home studio. It was a sweet setup with a beautiful board and perfect acoustics. Ted pointed to a big gelatinous black lump on a mixing console. He said, "This is the shit that put Bob in the hospital."

I was like, "What the fuck is that?" I had no idea.

He said, "It's heroin. You can shoot it up or smoke it."

He pulled a little piece of this black goo off the lump and stuck it to a four-inch square of tinfoil. He had a plastic straw in his mouth. He held the foil and with his other hand he lit a lighter underneath it, heating it up until the black substance bubbled. Thick wisps of smoke appeared. He inhaled the smoke through the straw and he handed it to me.

I did the same thing with the wad of heroin and, of course, was fearless. I took a hit and exhaled a massive amount of smoke. It gave me a small rush, and I felt a little light-headed. I was taking another hit when my stomach suddenly flipped. I tossed the heroin and ran to the bathroom. I could hear Ted's laughter fading out behind me. I knelt over the toilet and felt my head spinning. I was extremely nauseated and got so sick, I ended up puking all night. This shit did not appeal to me at all.

I had met a new girl a few days earlier and was seeing her from time to time. I called her, and she came over and played nurse. She really helped me out. She took care of me for the rest of the day in one of the bedrooms.

I talked to Izzy the next day and I was like, "Dude, I smoked this shit last night, I got so fucking sick." He said, "Where is it?" I said, "I dunno, I threw it down next to my drums." Before I could hang up the phone, Izzy was at the front door with some chick. "So where's that shit?" he asked. I pointed to the garage. They just marched right in and found the chunk lying next to a cymbal stand.

A couple of minutes later, while leaving to pick up some cigarettes, I spotted Izzy dragging the girl outside. I stopped the car and rolled down the window. "What the fuck's wrong with her?" I mean, *they just got there.* Izzy told me, "She'll be all right, she'll be all right." This chick was a rag doll, definitely not all right. She must've shot up and went out, like *bam.* He was dragging her out from the garage to get some air. That was the first time Izzy visited me at Bob's.

Hi. Where's the heroin? Bye.

A couple of days later, Izzy brought his gear over, and we jammed. I love Izzy. He defines cool and we really are good friends. He started coming by more often, and we played together all the time. He wasn't hitting the crack pipe at the time, but he'd face that demon later. Whether it was the presence of drugs or a new place to jam, it didn't matter to me. Izzy was just fun to hang with and I was happy to share this sweet situation.

## ⇥ WELCH GETS OUT ⇤

A few weeks after my setting up, Bob Welch was released from the hospital. When I met him he was a thin guy, bald on top. He had let what wispy hair he had grow long and fly freely. He wore a hat most of the time, flipped sideways, kind of French-beret style. I guess Ted had let him know about my staying at his house because Bob never had a problem with it and accepted me openly.

We would hang out and he would show me videos of some of his concerts, my own personal rock 'n' roll lessons, and he'd share all of his crazy stories. We watched tapes of him playing at the Cal Jam Festival in '74 where it's clear he's a fucking genius; he sang and played guitar brilliantly. Check out his stuff on Fleetwood Mac's *Bare Trees* and *Mystery to Me* albums. This was when the band still flexed its blues roots, and Bob penned and sang some incredible lyrics, among them "Hypnotized" and the aforementioned track "Sentimental Lady."

Although Bob had just gotten out of the hospital for overdosing, he was right back to smoking coke and heroin. To avoid these temptations and clear out my head a bit, I left for a day or two and couch-surfed on Sunset. It became clear what I had to do. I had gotten to that point (again) where I no longer recognized the guy in the mirror. When that happened, I knew it was time for a change of scenery. It was some internalized safety mechanism I benefited from at the time, but one day it would fail me when I needed it most.

I decided to get my stuff out of Bob's place and take off. I realized I had been there for two months, partying, smoking coke every day, and doing little else even with a tricked-out studio on the premises. When I came back to move out, Izzy was there, hitting the foil with Bob. He just gave me a slightly pained look as I gathered up my stuff, but it was cool between us.

Izzy had nothing to do with my having to get out of there, just as I had nothing to do with Izzy staying. It was like that with Duff, Slash, and Axl too; we all lived completely independent lives. None of us ever tried to will an agenda on the others. We each occupied our own orbital, doing our own thing, and eventually if it was meant to be, the cosmic debris would line up again.

# THE ORIGINAL LINEUP

It was time to regroup at my grandmother's place. That was always the spot where I could go to gather my sanity. It was my refuge, a simple place to take a shower and get something to eat. I had a backpack with some ratty clothes in it and I had my drums. That's it. I didn't wear underwear. I had three pairs of pants, some shirts. They were mostly concert T's: Aerosmith, Kiss, and Queen shirts. It was easy for me to pack up and go anywhere, anytime. I'm still pretty much like that. Possessions are chains that, over time, shackle and crush you. My drums and the shirt on my back; any other shit just slows you down.

After a few days back at Grandma's, I hooked up with the greatest job at a liquor store on Sunset and Doheney. I worked there for five months, earned six bucks an hour, and made great tips. The owner, Sid, was a fun guy. He was a freak, always partying hard. I had the Road Crew-zer car and would deliver alcohol, food, and cigarettes to his customers, mainly on large estates up in Beverly Hills.

Half the time I'd go inside and start making out with these older women. Often I'd end up drinking the booze that I brought them. We'd do coke and smoke bud. I'd have only three or four orders, which the average driver would knock out quickly, but not me. I'd usually find my way back to the store a couple of hours later.

I delivered to this one hot woman, Laurie, who had worked at the Rainbow for like twenty years. She was a classy lady, a real pretty blonde. She lived right behind the store. I brought stuff up to her nearly every night. She was a real sweetheart. Looking back, I became friendly with just about everyone on the delivery route.

I'd go to Sid's house sometimes. It was a nice, big place, plastered with pictures of him with big stars like Frank Sinatra. He had this safe in the floor under his bed, concealed by a carpet. I remember he opened it one time and I saw a big bottle of quaaludes, a bag of what looked like kick-ass smoke, and a big wad of hundred-dollar bills. He liked boys, so he always had a few of his young fellows up there walking around in their little bathing suits.

He would tease me sexually, gentle come-ons, but no way I was going to ever be one of his playthings. And I was never offended by anything he said. He knew I didn't swing that way and that was it. He was a real character, an aging bald guy who reminded me of the classic movie actor Edward G. Robinson, a fleshy round man who had done well and couldn't have cared less about losing his looks and being out of shape. He worked hard in his early years, and that store meant he was set forever.

## ⇥ FIRED AGAIN ⇤

When I got fired from there, I couldn't believe it. I got back from a delivery on a sunny Sunday afternoon, when it wasn't too busy. I was stoned as usual and totally parched. I went into the cooler, because as the delivery guy, I also stocked the store. While I was in there I poured a small amount of wine cooler into a plastic cup and drank it down. That made me feel better. I popped out of the cooler and walked up to the cashier. "Hey. Any calls, anything happening?"

He grabbed the cup out of my hand, took a sniff, and said, "That's alcohol!"

I looked at him and said, "*Yeah . . . ?*" It wasn't the first time I had booze breath on the job for chrissakes.

He looked at me like I just shot his mom, picked up the phone, and called Sid. Sid had no choice but to fire me, because it was clear that this nerd-detective cashier was determined to make a big deal out of it. I kind of understood. Sid just didn't need the hassle. But I was really bummed to lose that job.

Within a couple of days, however, it was life as usual. I would just run the streets with these goofy young girls I'd meet at the exotic dance clubs. They were strippers, so they had money and a place of their own. I always had a place to crash, but I never fell for any of them; they were just fun to party with. I'd bump into Slash from time to time, but he was currently in a pretty serious relationship with that raven-haired hottie. During this entire time we weren't playing together much, just loving life. And it wasn't like we missed playing. The time would come when we felt like jamming again, and time was on our side.

## DUFF'S THE STUFF

In the fall of 1984, I scored another very decent gig. I moved in with a bass player and a guitar player, Jeff and Todd. Jeff played guitar and lived with his father, who owned a nice four-bedroom house in Granada Hills. Just playing with them was cool enough, but I had a bedroom there too. The garage was a converted studio where I just played drums with these guys and was able to practice every day and have a free place to live. I couldn't have been happier.

I had all I needed. But the band failed to score any gigs. Nobody took the initiative. After about two or three months, it was time to move on. All I wanted to do was get out and jam as much as possible with other musicians. I began to feel frustrated and every day I felt a growing desire to play.

If I wanted to get good at it, I knew what I had to do. I got together with Slash again, and we decided to make our often discussed but not yet materialized Road Crew project happen. First we enlisted our old friend Ronnie Schneider on bass. He had a cool image, played well, and was my longtime bud-puffing buddy. There's a picture of Ronnie in the *Live Era* album; he's playing guitar next to Slash. He's the one with the blond hair that's all teased out.

Ronnie soon left to join another band, so we took out an ad in the *Recycler* under "Musicians Wanted." We got a call from this guy who said he was from Seattle. He explained that he used to play the drums in a punk band, but since there were so many drummers in L.A., he took up the bass. It cost money to rent a studio so we figured we should save our money and just meet him first to see if he was a cool guy to begin with and what kind of look he had. Image was important, and if you thought someone looked cool and could play at all decently, that nailed it. You took him on. If he didn't look the part . . . next.

We decided to meet him at what I personally consider to be the all-time legendary Guns N' Roses landmark: Canter's Deli on Fairfax. Slash knew the owner's son, Marc Canter. He's our age, and he runs the place now. Slash and I would go there every day to get meat knishes with gravy for a dollar, really tasty. The picture on the back cover of *Live!? Like a Suicide!* was taken in the alley behind Canter's.

Mark ended up taking pictures of the band at our shows. He was a smart, artistic, compassionate guy. We felt comfortable trusting him to shoot all the behind-the-scenes images of GNR. We knew he wouldn't compromise our trust, wouldn't sell out to some rag-ass tabloid, or let anything out that we didn't approve. Mark and I are still close to this very day.

## ⇥ BUILDING THE BAND ⇤

So we're hanging at Canter's on a sunny afternoon in the winter of '84, and in walks Duff McKagan. Right away I thought, "Well, he sure as hell looks like a rock star." He was tall, six feet four, and had long, teased blond hair with a black streak running down the side. He had been calling himself Duff Rose and was all rocked out. I thought he was totally right for the look we were after. We hit it off right away, right down to the bands we liked, and just as important, the bands we hated.

Although we got along, we didn't actually get together to play. We did hang out a bit, and during that time, we introduced Duff to Izzy, but nothing really got going. For a good amount of time there it seemed, once again, as though putting together a band wasn't going to happen. There just seemed to be a lack of drive or purpose.

After admitting to myself that the Road Crew idea had fizzled out for a second time, I started having these depressing thoughts about what the hell I was going to do. College was out, and my gut still told me that being in a rock 'n' roll band was my best bet.

I was feeling so down about nothing happening that I dropped in on the local navy recruiter. I figured what the hell, I had nothing to lose. See the world on someone else's dime. I took the test, and the guy there told me they'd get back to me in a few days. He shook my hand and smiled. Leaving the place I felt that I would soon be setting out on a whole new adventure.

Well, I must have sucked on the test, because I never heard back from them. Admiral Adler was not setting sail in this man's navy. Sometimes I shudder when I think that I could have been swabbing the deck on some boat in the North Sea instead of touring with GNR.

## ⌁ EMPLOYED AND BUSTED! ⌁

When I was nineteen, I found a job working for a computer chip company in a warehouse in Chatsworth. I was a packer and a shipper, making $5.35 an hour. The secretary there was a dirty-blond hottie in her late twenties. She and I would go to lunch together every day at Chatsworth Park. I would do her in her backseat during our break. Love those power munches—er, lunches. She partied and sold a little weed on the side, so we would sit under a tree in the park and take rips off of her bong. One day while we were getting stoned, these two guys came walking toward us. They were about twenty yards away when she covered the bong and the weed with her hat. They walked right up to us and flashed a badge. I was like, "Fuck!"

One of the cops lifted her hat, exposing the multiple bags of weed. They arrested her and, for some reason, just let me go. But of course, word got back to our employer so I could extend my streak of getting fired from every damn job.

## ⊰ STAYING STOKED ⊱

Getting canned didn't upset me that much because ultimately all I really wanted was to be in a band, and playing the drums was the one job I took very seriously. I would drive anywhere, meet anytime, sit down with anyone just so I could play. I met a bass player named Gary who had a recording studio out in Burbank, so I started jamming with him. Bobby Chouinard, the drummer for Billy Squier, played on one of Gary's demos, which was great because he had his very own recording studio and tapes of his fully produced music.

Bobby's studio was in Laurel Canyon, up in the Hollywood Hills. I felt like I was finally making some headway with the music, performing and listening carefully to the playback, picking out my strengths and what I needed to work on. Of course we were partying too. These guys loved to smoke weed and jam. It was a great vibe all around, solid musicians, great studio, kick-ass weed.

## ⊰ GNR CYCLES AROUND ⊱

I played with them about a month or so and then one night Slash called me up. He sounded excited and told me that Izzy had resurfaced and wanted us all to play together again. Like I said, that's how it worked back then. Things just had to take their own natural course. Something would pop up, a booking, a festival, a fucking keg party, and a few phone calls later, we'd be getting together again. This time, however, my heart really started pounding because Slash told me that they had committed to doing a show Thursday night. And Friday they were planning on heading up to Seattle to play a couple of shows. Since we had introduced Duff to Izzy and Axl, he had been playing with them too. In fact, Duff was the one who booked the upcoming shows.

So it was "Dude. Cool. I'm there!" The next day I got together with them, and they told me the band was now called Guns N' Roses, after the band's founders: Tracii Guns and Axl Rose. So I guess technically, the very first Guns N' Roses lineup consisted of Rob Gardner on drums, Tracii Guns on lead guitar, Izzy on rhythm guitar, and Axl.

Tracii was someone Slash and I knew from Bancroft Junior High School. He used to have the total surfer-boy look with the straight blond hair that all the girls liked. Now he sported jet-black hair and tattoos. We didn't really hang with him that much and never jammed with him. Axl and Tracii had a place together, and it was there that they came up with the name for their band.

For one reason or another, however, Tracii and Rob weren't up for the trip to Seattle. I guess it wasn't hard to see why they felt that way. The thought of humping their asses nearly a thousand miles north for a couple of gigs that wouldn't even cover gas money probably struck them as pretty fucking dumb.

So Izzy called Slash and Slash called me, and for us it was "Hell yeah!" from the start. It wasn't even something I had to think about. I loved playing in a band and I loved rock 'n' roll so much that it was a no-brainer: "Where's the gig? Siberia? Okay, I'm there." Simple. After that epic road trip, we were pretty much inseparable and became the founding members, the classic *Appetite for Destruction* lineup.

## ⚔ MY DRUM SETUP—THE TRUTH ⚔

The Thursday-night gig, before we headed north, was at the Troubadour. So we practiced Tuesday night and Wednesday. I still remember what it felt like when Slash and I walked in to set up for that Tuesday practice. There was zero awkwardness, everybody got along really well, and there was a distinct willingness to make it happen. To Izzy's and Axl's credit, there was complete respect for what we were each bringing to the table and none of that newcomers-versus-veterans bullshit. We jammed to a Stones song, "Jumpin' Jack Flash," and an Elvis oldie, "Heartbreak Hotel," and then kind of looked at one another. I definitely had a sense that something special was brewing.

I honestly don't think it was just me who felt that way, because Thursday night, something permeated the show at the Troubadour and it went pretty well. I remember we played for only about ten people, and it didn't matter. We were playing for the music, for the sheer excitement of performing live. There's actually a photo from it in the *Live Era* record, and you can see Michelle Young standing to the side with her hand in the air.

That was also the first time in forever that I played with a single bass drum. And you know what? I loved it. Ever since that show, I decided to keep my drums set up that way. Now, maybe you've heard all about how Duff and Slash hid one of my bass drums on me before the show, forcing me to play with just one bass in my setup. And there's a story out there that they knew in their hearts if they could just get me to play one gig that way, I'd be sold and never use a double bass again.

It's a sweet little story, but at the time, the only thing Slash hid on me was his stash. The simple truth is that one of my bass drums was busted. Somebody had fucked it up by dropping it or stacking an amp on top of it. So I had to go on that night with a setup consisting of a bass, a snare, a floor tom, one ride, a crash, a high hat, and a cowbell. Necessity can be a motherfucker, but that night she did me right.

Now, we must have been deadly serious and not a little nervous about our first road trip the next day, because I don't remember going out and partying that hard after the Troubadour show. Suddenly, it was Friday morning (late morning, natch) and the time had come to drive up to Seattle. We loaded all our shit into our friend Jo Jo's car. Jo Jo was this raw-lookin', rough-edged kind of guy with long stringy brown hair. His brother Raz was a buddy of ours too. Raz was confined to a wheelchair, and a greater, more enthusiastic soul you've never met. They were willing to do anything to help us out and were both really great guys.

Our friend Danny was coming along to give us a hand too. He was a cool kid with short blond hair that he had all spiked out with gel and hairspray. They all did their part and roadied up for us. We were all the same age, and we all busted ass to get the show on the road. There were no chiefs and no peons; we were all equals, all like brothers. We were on our way. Eight hundred forty-five miles to our first paying gig on the road. This was it, damn the torpedoes, no looking back.

We got as far as Bakersfield.

Our shit box of a car, I think it was an Oldsmobile, just died on us. We couldn't believe it. But we decided we were going to get to the gig by any means necessary. Danny and Jo Jo agreed to stay with the car, get it fixed, keep an eye on our gear, and somehow get it all up there. I grabbed my stick bag, the guys grabbed their axes, and we started walking along the freeway with our thumbs in the air. Have Guns N' Roses, will travel.

Five long-haired, cocky punks, each one in the best of moods, set out to fulfill their destiny. We hitched a bit, then while the other guys were taking a break sitting on the side of the road, I managed to snag a ride from an immense eighteen-wheeler. We all piled in and were able to get as far as a truck stop just outside Medford, Oregon. There, a Mexican farmer and his son stopped for us in a ratty pickup truck, and we all piled in the back. Unfortunately, we were way too heavy, and the tires started rubbing against the fender wells. We pretended not to notice that there was smoke everywhere (we'd rather have suffocated from the fumes than gotten back out and walked), but naturally, he couldn't take us any farther. We thanked him and got out.

About an hour later, two wild hippie chicks blew by us as we waved our thumbs in the air. I yelled, "Shit!" But I could see they were checking us out as they flew by. I crossed my fingers and watched as they made a lazy U-turn, came back, and picked us up.

They both had waist-length hair and were dressed in colorful commune clothes. We put our guitars in the trunk and piled in. They said when they were younger, they used to hitchhike everywhere and would get pissed when no one would pick *them* up. That's why they came back for us.

They were totally sweet women with some sweet weed. I remember chatting with them: "Oh, you lovely ladies. My kind of girls." I was in the backseat leaning forward with my arms around them. They took us to Portland. Then Duff's friend Greg drove down from Seattle, picked us up, and took us to the Gorilla Gardens, the filthy dive bar where we were to do the show that night.

When we got there, we walked right onto the stage, and just in the nick of time. We didn't have time to grab a beer, smoke a joint, or put on makeup. Although we were still at the stage where we'd tease our hair up to God and slap on the eye shadow, heavy eyeliner, and lipstick for our stage performances, there just wasn't a minute to spare. Later I'll get into how I put an end to the whole tedious makeup routine. Most fortunately, we were able to use the previous band's equipment. We just went on, jacked in, and played our songs.

We opened with "Reckless Life," then did "Shadow of Your Love," "Move to the City," "Anything Goes," and "Jumpin' Jack Flash," and closed with "Heartbreak Hotel," which was Axl's choice. Axl often warmed up his voice with an Elvis cover and was a big Elvis fan going back to his days growing up in the Midwest. "Heartbreak" got the biggest response; "thankyouverymuch," King. And although we didn't exactly bring down the house, we got decent applause and were all smiles after the show, feeling that for the most part, it went over pretty well.

Afterward we had drinks with Duff's friends, and a nicer, more gracious bunch of partiers you'd be hard-pressed to find. Duff was very popular around the Seattle area; everyone knew and liked him. His buddy Greg invited us back to his place. I walked in the house and I was like, "Oh, yeah." I was in heaven. He had at least fifty plants growing in his basement, the best pot I had ever smelled, and at the time, it was the best weed I'd ever toked. Greg, his girlfriend Jill, and I just smoked out.

Greg's girlfriend made us a big spaghetti dinner, and then we smoked some more. The next day she gave us a ride, not to the state line, not to San Francisco, but all the way back to L.A. Thank you, Jill, that was so incredibly cool of you.

When we hit town, word got back to us through Duff's network of friends that Jo Jo and Danny had fixed their car. The drums and the gear were safe. All in all, it was a miracle. We had hitchhiked up the whole West Coast with no gear and no money, and although we missed the other two gigs Duff had lined up, we pulled off the main booking in Seattle and not one of us ever complained.

That trip was the acid test, the mondo bondo epoxy that sealed our fate as the original Guns N' Roses. Ever since that trip we stuck together like we were one creature. It was as if we had set up this ridiculously irrational initiation ceremony, one that no sane person could have tolerated. Because that's what GNR had to be all about. We *never* did anything the sane, sensible way. We never went by the rules and never conformed to an accepted path to success. The way we came up with our songs, insisted on total artistic freedom, the way we practiced and played—no one did it like we did.

It was "Break the mold to make the mold." We all fed upon the same primal gut drive to take that road trip. After that experience, we knew that we were the only guys on the planet who could make this band happen!

# GROWING PAINS

≼ BUILDING THE BAND ≽

O ur behavior was to foster an "us against them" attitude, and that approach served GNR well, driving us to make rock with an immediacy and ferocity that no one had ever attained. Once back in L.A., we found ourselves totally motivated. We really bore down and practiced all the time. We started rehearsing at this guy Nikki B's place. His house was by the L.A. zoo. It was a dumpy dwelling in an industrial area literally plopped in the middle of no-where. There was a junkyard on one side, and on the other side was a big warehouse under construction. Danny or Jo Jo would drive us to practice every day. That was our rehearsal spot for a while. Then Nikki B joined Tracii Guns in his new band, L.A. Guns, and we had to find another place to jam.

Our minds were blown when they recruited singer Phil Lewis. Phil was a Brit who was in a band called Girl. They had two albums out in the earlier eighties, and their guitarist, Phil Collen, went on to join Def Leppard. Things were definitely getting hot for Tracii. And you know what? Good for Tracii.

After we returned from the Seattle trip, our first show was again at the Troubadour. Shortly after that we played Madame Wong's. With Nikki B's no longer an option, we started rehearsing at a studio behind the Guitar Center on the Sunset Strip.

That's where Slash came up with what we all thought was this awesome riff. He said he created it to limber up his fingers, get them loose before playing. He sort of made fun of it, saying that in his head it sounded like the notes you'd play for circus music, the kind you hear on one of those tinny pipe organs. If you've ever listened to the organ opening on George Harrison's "It's Johnny's Birthday," you know the sound I'm talking about.

I told Slash he was overlooking the enormous potential of that lick: "That's a great fucking riff, dude. We have to figure out a way to get that into a song." Artists have taken segments of music only meant for limbering up and transformed them into hit songs. Edgar Winter did it with a simple percussion exercise that ended up becoming his hit "Frankenstein."

So Slash molded the riff, and today we know it as the intro for "Sweet Child O' Mine." What I loved was that Slash truly displayed his brilliance by not just using it as the intro but finding a way to thread that riff throughout, using it as the backbone of the entire song.

At this point, however, we could count on one hand the number of rehearsals Axl had been to. He didn't have a PA system back then so he never went to the studio to sing. Sometimes, he would sit just outside the studio door and sing along, but usually we would just give him a tape of our rehearsals and he would go off with it somewhere. Many times we would do a show without any idea how Axl was going to sing on one of our new songs. We'd been around, and we'd never heard of another group that could operate this way. But like I said, it was becoming more and more obvious that GNR didn't do things like other bands, and the birth of "Sweet Child" was just one example.

Later, Axl told me that when he first heard the "Sweet Child" riff, he didn't need to be in the same room with us; he could have heard it over a phone on the other side of the globe. He'd listen to a cassette over and over again until he worked it way down to the marrow. He wrote lyrics in his little hovel upstairs and actually preferred it that way. Axl really seemed to like keeping to himself, not because he was stuck-up, or shy, or because he needed a better PA system, but because it was his way, his own thing. We managed to see enough of each other and were playing gigs two to three times a month at that point.

After we got back from Seattle, Duff, Slash, and I started hanging out all the time. Axl was a loner who wrote killer lyrics about who we were and how we lived and what we were experiencing at the time, and Izzy, well, "Izzy's just Izzy," I'd say, and we'd all nod. Izzy ended up popping in about half the time. Again, there was no pattern, no agenda.

## ⇥ THE HELL HOUSE ⇤

We had adopted a permanent hangout. We called it the Hell House, as it was this old dilapidated shack occupied by Axl's friends West Arkeen and Del James, a biker. It was located at Santa Monica and Poinsettia, right by Gardner Street. We rented it for around $600 a month. Well, we never actually paid a dime, but somebody must have been chipping in.

Now, West Arkeen was a real character. Axl hooked up with him through some chick. It was as if everybody met somebody through a chick they were fucking, wanted to fuck, or had fucked. That's how we ended up at the Hell House, through a fun-loving, cross-pollinated chain of people. We were going there every day, so we agreed that since we were there all the time, we might as well make maximum use of it. Hell House became the band's official headquarters.

We hung out, partied, puked, and passed out there. It was our preferred crash pad. It was there that Axl had his friend Bill Engell design a now famous tattoo for him, a cross that featured five skulls, caricatures of each member of the band. Slash gave Bill a hard time because he couldn't re-create Slash's curly fro and Slash's skull ended up having straight hair.

The regulars at the Hell House included Duff, Slash, Izzy most nights, and me. Jo Jo, Raz, Danny, Dizzy, Del, and West were there almost all the time. Axl liked writing songs with West. He liked kicking it while West played guitar. I was there all the time, literally spending whole days and nights. There were always random people crashed out on the floor. It was a never-ending revolving door of derelicts, a hilarious party scene.

Out of this drunken wasteland everyone kind of spontaneously formed a fun jam band called the Drunk Fux. Many different people were in that band, including Tommy Lee and Lemmy. It was just a jam thing really, and we played some free benefit shows around L.A. Maybe one day we can get Mötley Crüe, Mötorhead, and GNR to reunite the Drunk Fux, the ultimate superband of the eighties.

Axl, West, and Del had their own little clique that wasn't really part of the Drunk Fux, and I couldn't have given less of a fuck about it. I don't mean that as a slight to Axl. I just wasn't into pining away at not being asked into his elite crew. I got along with everyone and was always laughing, having the time of my life.

## ⊰ AXHOLE ⊱

always thought Axl was a totally cool asshole. I knew that he was a fucking star, a truly great performer. But I was also aware that at times, he could be an insecure prick. As long as he wasn't fucking with me, however, we were cool. That's how it was. Then he pulled the first of a series of fucked-up shit that he did to me over the years.

I remember Axl was staying with Jo Jo at his apartment. I stopped by to hang out a bit. I just opened the door and Axl jumped up and lunged at me. The place wasn't that big so he only had to take two steps.

It happened so fast, I was like, "Huh?" He hauled off and kicked me in the balls. I could tolerate a lot of bullshit from Axl because he had some really unfortunate hang-ups, but getting my nuts cracked was the last thing I expected. I doubled over from the pain, and my eyes teared up. Then, when I was finally able to breathe, I just yelled, "Fuck you!" and left. It was the weirdest goddamn thing. But ultimately I let it go. At the time I felt I had to.

My older brother, Kenny, would do shit like that to me too growing up, so I didn't take it too much to heart. I went back to Izzy's place and told him about it. He was surprised and just said, "I don't know, dude." That was the law of the Axl; you never knew why. I never did anything against him. Any chick he liked I wouldn't fuck, although some made it clear they wanted me. If Axl was interested, I figured it was his girl. I could respect that because in the end, I didn't care and everyone knew how insanely fucked up he was around women.

I became more frustrated with Axl's actions over the next year. Axl's behavior became seriously unpredictable. He was getting into fights, often starting shit at the Hell House with random people who came by to party, so they just learned to give him lots of room. Some of the uglier incidents were just hushed up, because, well, it was Axl. Axl had only one rule for himself: there are no rules.

Duff loved walking into some random club just as everyone was looking for a party or something to do after-hours. He'd invite them all back to the Hell House until they were pouring out into the streets. We would just shake it up and by four A.M., there would be a hundred people milling around.

And we were pretty resourceful once the girls got there. When one of the guys would be fucking some chick, one of the other guys would go through her purse and take like five or ten bucks. We would never steal all her money, just a small amount, because we really needed it. But the Hell House could have also been called the Shit House. After a year of constant abuse . . . well, you get the picture. Plus, the cops were starting to come by all the time. By the second year, more than half the parties we had there were busted up.

So I went into my need-a-little-space, time-to-regroup-and-preserve-my-sanity mode again. Across the street from the Hell House was a duplex, which had two little apartments on one floor. My friends Julie and Tracey lived there. Julie would later appear in our "Welcome to the Jungle" video. She's the girl who catches Axl's attention in the beginning, wearing fishnet stockings walking down the city street. She's also seen lying next to me on the bed.

So I rented a room from her, and talk about a totally cool landlord. The room was just their washroom, where there was an outlet, a splash sink, and hot and cold water faucets where you could hook up a washer and dryer. I put a piece of wood over the sink, and I put my TV on top of there. I even tapped into the building's cable, which was kind of cool for me because I never had cable before.

I had a futon mattress that didn't fit flat riding up on part of the wall. I had a blanket and a couple of pillows. There was a private door on the side where the driveway was. I'd go in and out whenever I wanted to. But more important I had a steady place to get away, sleep, and fuck.

There were three girls I loved having over regularly. One in particular, Adriana Smith, was a hot Native American chick with a hard body and such a pretty face. Izzy introduced me to her and we became close. She was in a gang of crazy biker chicks.

Then there was Gabby, or "GabaGabaHey." She was a short, hot rocker who could jam all night. And finally Adriana Barbour, a cute, timid Valley girl who could drink any man under the table. They worked at the Seventh Veil gentlemen's club on Sunset. Their apartment was right above a pool where we would get so fucking drunk. We named ourselves the Naked Skydivers from Hell, and we would jump from the balcony into the pool. We somehow avoided getting killed during one of our many insane diving stunts.

## ⤞ PRACTICE MAKES PERVERTS ⤝

GNR continued practicing in the space behind the Guitar Center. Jane's Addiction rehearsed in the same studio, and I would see them every day, because I would always set up early while they would be breaking down. We had parties there too, because we would always end up at the studio after kicking it at the clubs. There were so many girls who wanted to hang out and be with the band.

Slash and I would get these girls and fuck them back and forth.

It became sex everywhere, anywhere, anytime. It just happened. If I saw a girl I liked, I'd approach her, and that would be it. "You and me, right now." There was no such thing as warming up to the situation. The situation was "Hey, bend over the washbasin!"

Girls also knew to show up at the studio. I remember one time Izzy and I got a blow job from some girl after one of our shows. It was about two A.M. She sucked one of us off while she jacked off the other, and then she would alternate. We were outside by the back door looking up at the sky going, "Oh my, oh fuck, oh shit." She swallowed Izzy's load at the moment I came, and I shot all over her face. I couldn't leave her like that, so I went in and grabbed some tissues so she could clean up. Shit like that happened all the time.

## BIG LILLY ROCKS

The band developed a strong following early on, and over the years, these fans would prove to be very loyal. They embraced our reputation as badasses who tore up everything in our way, because that's exactly what we were. GNR was a fucking whack-job gang that was ready to rumble. We would fight tooth and nail to get a gig. We started playing the Troubadour on Tuesdays, then we moved to Thursdays, and then we began playing weekends as headliners. This all happened pretty quickly, because we never stopped bugging them about booking us better times and days.

My mom went to a few of our gigs. She really liked being a part of it, and Jamie loved going to the shows too. It would always be a special night when I saw members of my family in the audience. Even my grandmother made it out with Mom and Jamie one night to see me play at one of our shows at the Troubadour. The place was packed.

My mom told her, "Stay right here, I'm going to get a soda." When she left, my grandmother, feisty little groupie that she was, pushed herself way up to the front of the stage. The show was a sellout, and I remember looking down and seeing her with a big smile on her face pointing at me and telling anyone who would listen,

"That's my grandson, that's my grandson." I never saw her looking so proud.

As fate would have it, it was one of our performances where Axl was wearing his black G-string with his bottomless chaps. He was dancing around shaking his ass right in front of her. I was cracking up, yelling to my grandma, "Get out of there, Grandma. Better clear out of there!"

In addition to the Troubadour, there were two Madame Wong's nightclubs, Madame Wong's East, downtown, and West, in Santa Monica. When we played there we would draw anywhere from twenty-five to a hundred people. The Roxy was a bit more prestigious, and you really had to hustle to bring in people at the Roxy.

After the Whisky reopened (it had been closed for a year from '84 to '85), we played the second night of its reopening, and it was packed. It felt so cool to play a famous landmark.

## ⊰ BOTTLE ROCKET ⊱

Every year in L.A. they held what was called the Street Scene. There were ten or more stages set up, all featuring free shows. It took up a few city blocks, and by the time we were asked to participate in 1985, it drew about a hundred thousand people. We were pretty familiar with the festival and felt that the gig could potentially get us some good exposure.

Unfortunately, it turned out to be a very fucked-up show. I was on a stage setting up my drums, putting the bass drum in place. All of a sudden, this empty Jack Daniel's bottle comes *flying* past my face and nails my cowbell. It missed my head by an inch! Some dumbass really tried to hurt me.

During our set, people were actually *spitting* at us. It was pretty ugly. I think this was some sick remnant of the masochism and self-abuse of the punk era. It was odd, and it was dangerous, but most of all it was sad; the punk kids were in the death throes of a has-been era. Shove a nail through your dick and parade it in front of your friends, because, well, punk is fucking lame.

We didn't stand for shit like that and the band was spitting right back at them. It got so ridiculous it became funny. I remember seeing Duff looking all pissed as he hocked a big loogie into the crowd. It definitely was an unforgettable performance.

Poison played the Street Scene that night too. We saw them play clubs, get signed, and become huge. They started out about a year before us in 1984. I would help Rikki with his drums from time to time. They had this really cool guitar player, before C. C. Deville, named Matt. I was hanging out with them one day, and they asked me if I could run the spotlight for them for a show at the Country Club. Why, sure.

They told me that whenever Matt did a solo, I should put the spotlight on him. So I was up on the balcony running the spot and I have the light smack on Matt during a guitar solo. During this, Bret goes dancing off the side of the stage. After Matt's solo, I'm maneuvering the spotlight back around in search of Bret. Bret wasn't singing anymore. He was nowhere to be found. The guys in the band looked bewildered. Next thing you know, someone's calling for an ambulance. Bret had fallen off the stage. There was a big square hole on the side of the stage used for storage that wasn't clearly visible during the show. When Bret toppled in he broke three ribs, so they had to stop playing, and the show was over. I felt so bad and wished I could have prevented that from happening.

The four guys in Poison shared a one-bedroom apartment. It was disgusting. Now, for me to say it was disgusting, you know it had to be pretty sick. The place was crawling with cockroaches. They had four queen-size mattresses set up in the bedroom, so the floor was like one big mattress. They had chicks there all the time. Poison had girls bringing them food and money and they were all hot. So when I saw what they had going on I thought, "Oh yeah, this is going to start happening with us."

Matt was my buddy. He always invited me along with the band to gigs. He ended up leaving Poison in 1985, and I was like, "You're crazy. What's wrong with you? You guys are gonna make it big." But he was done with it. He just wanted to go home and get married, make babies, and ride into the sunset.

Matt faded out about the time GNR came back from our first trip to Seattle. With Poison's guitar spot now open, Slash went to them for an audition. They acknowledged Slash's kick-ass ability, but they didn't think he was pretty enough to fit in with their glam image. They had their own image to maintain, their own way of doing things, and you had to respect that.

In fact, whenever Poison played it was tough shit for whatever band was going on after them. Bret would announce to the crowd that they were all invited to an after-hours party and hand out flyers with directions to their apartment. By the time the next band was on, the place was nearly empty.

I thought Poison was just a cool rock 'n' roll band that was doing exactly what I wanted to do. I remember Bret and Rikki were at one of our shows at the Troubadour. We could see them standing right in front of the stage, a few feet back. At times, Axl could get mean talking to the crowd, and that night he said, "Some guys from another band are here tonight. Don't they know that nice boys don't play rock 'n' roll?" That was our cue to launch into the song. But he said it so seriously that night, it was an obvious jab at Poison. They didn't give a shit though; they were the top band at that time. In comparison, I could not imagine Bret ever being a dick like that to anyone.

# RULING *the* STRIP

⊰ FEEL THE LOVE ⊱

When GNR started getting popular, we had lots of chicks hanging around us all the time. And just like with Poison, they were making us meals and giving us money. I fucked them all, fat chicks, skinny chicks, plain chicks, shy chicks—it didn't matter. I was just showing them my appreciation.

It got to the point where there would be lines around the block wherever we played. I heard so many club owners tell me that we were going to be huge; they hadn't seen such crowds turning out for a local band since the rise of Mötley Crüe. During our show there would be hundreds of camera flashes going off, and while the guys clearly dug us, the fucking babes would become hysterical. The drummer always has the best seat in the house, and at every show I'd notice hotties shove their way to the front and scream nonstop. That was the clearest indication that we would make it, and I never once doubted it.

I'll never forget one of our shows at the Troubadour. I was getting ready to go out and play. My hair was teased up to the fucking ceiling, and I had on a sleeveless ruffled white tuxedo shirt that I had tucked into my black leathers. There are these windows in the dressing room, and I was sitting out on the ledge of one of them, smoking a cig, looking at the marquee with "Guns N' Roses" spelled out in big block letters.

Down on the street level was a long line of people waiting to get in. Some fans looked up and saw me. They yelled, "Guns N' Roses rules!" Everyone looked up and cheered. A thrill ran through my body like I had never experienced before in my entire life. I was a part of something that felt incredibly exciting, something that got people whipped into a frenzy. Something I had helped to create was taking on a life force all its own.

To this day, I get the same rush when I drive by the Troubadour, something no one can ever take away from me. It's one of the last places on earth I'm guaranteed a natural high to this day. That night I knew that exciting things were about to happen and the burn in my gut told me it was going to be huge.

In 1985 the Sunset Strip was like a nuclear reactor approaching critical mass. Heavy metal bands flooded the scene, and since there weren't enough clubs to accommodate all of them, a pay-to-play rule was put into effect in the most popular spots. This was basically an insurance policy put in place by the club owners. If a band wanted to play a certain spot, they would have to buy a minimum number of tickets off the owner, and then it was up to the band to sell them.

This amounted to a pretty shitty deal for the bands. And as far as I can remember, we never agreed to this. We would never buy into their pay-to-play policy. Even in the beginning when we'd have only like fifteen fans show up, we never paid for anybody. I remember from time to time I'd wind up with a stack of tickets shoved into my jacket. But I was like, "Fuck this." I didn't sell any. I really don't think the other guys did either. But I think one of our more enterprising techs used the money he got from the tickets to buy drugs. Now, why didn't I think of that?

We were unaware of it at the time, but we were carving our own path to glory and it was about to look as well as sound like nothing else out there. At first our look just followed along with the glam scene. We all wore makeup and teased our hair up to the clouds. A look that had started innocently enough with Marc Bolan of T. Rex had gotten way out of control, and we didn't even think about it. But the whole glam look could take over an hour of preparation before a show, and something had to give. I am proud to say that I'm the one who put an end to the pretty-boy bullshit.

It happened one night after a steamy show at the Troubadour. I came backstage before the encore and couldn't take it anymore. I felt like I was suffocating and honestly wanted to jump out of my skin. I grabbed a towel off the counter and rubbed my face raw.

The boys looked at me. "What are you doing, Stevie? We got to go back out!" I exploded and yelled that I was done with the whole makeup bullshit. I just couldn't bear it anymore. Drummers sweat the most and the shit would run all over, down my neck and chest and onto my clothes. The worst was getting makeup in my eyes. It burned like a mother. Plus, I've got body hair like fucking Wolverine, so I was already running hot out there and enough was enough!

I tossed the towel over a chair, and Izzy pointed to it. It had this bizarre imprint of my face on it. It was like that story where Veronica, some random woman who took pity on Christ, wiped his face while he was carrying the cross, and all his blood and sweat left a lasting impression of his features on it. Or maybe I'm confusing it with the sheet they wrapped around him when he died. Anyway, I think it's still around today in a church somewhere in Italy. The Shroud of Turin. For me it was the Shroud of Touring.

That was the last time I wore makeup. You know what? By the next show, I looked around and the war paint was gone from all our faces. We were so over the eighties makeup scene and we all realized it wasn't us. So good-bye to glam.

It was our sound that really mattered and set us apart anyway, and losing the makeup kind of accentuated that fact. Nobody was doing the kind of white-hot blistering pure rock that we were cranking out. A lot of our competition featured whammy-bar pseudo-Eddie guitar gods who had to have both hands ripping over the frets. I swear, every other band on the Strip featured these Van Halen/Randy Rhoads wannabes. They were fronted by castrated falsetto screamers like Queensrÿche's Geoff Tate.

Slash didn't go in for all that fancy guitar wizardry, and Izzy absolutely despised it. Izzy embraced the no-frills power chords that lent to the infectious gut-punch rhythms of a Keith Richards or Pete Townshend. Slash idolized Aerosmith's Joe Perry, who combined incredible chord work with impeccable solos. So it was during this time that GNR evolved into a faithful rock 'n' roll outfit. Duff and I laid down the foundation and Izzy and Slash built their masterpiece on top.

## ⇥ ADLER TO AXL ⇥

While other bands were singing about dungeons, wizards, and black magic or partying in the backseat of Daddy's car, Axl was writing lyrics about life, his life and our lives. We all had lived through some pretty dark, twisted shit, but it was real, and the kids sensed it and responded. Axl excelled at capturing the mood, and whether it was "Nightrain" or "November Rain," there wasn't a soul in the audience who couldn't feel what we were shouting about or living through. "It's So Easy" also ended up being a song about our lives in the now, at that very moment. And nobody in the world sang with more intensity, more honesty, than Axl.

GNR just played the kind of rock 'n' roll that everyone loved. We were the wanton offspring of Aerosmith and the Rolling Stones, delivering the goods with a hard-core rock attitude. We were quickly recognized as the seediest primal band in existence. The few bands that were similar to our flush-the-fashion, don't-give-a-damn look were also bands that often shared the same bill as us: Junkyard, Faster Pussycat, and the Joneses. The Joneses were these hip-looking street rockers who didn't go out of their way to be too glam. Duff was close with them. There was also a cool chick band called Hardly Dangerous that Axl liked to have around. They were hot.

We also hung out with Taime Downe from Faster Pussycat. He started the Cathouse with another one of our friends, Riki Rachtman, who was in a band called Virgin. Their logo featured a cherry with a bloody knife going through it. It was quirky, whack, but funny stuff.

He and Taime really had the Cathouse jumping, and it became an extremely popular and successful rock club, with Guns N' Roses playing no small part in its taking off. In fact, the Cathouse became Guns N' Roses' own personal hangout, where the DJ played our songs in a club for the first time.

## ⌐ THE CATHOUSE ⌐

The Cathouse started off over at the old Osco's Disco building on La Cienega, where Slash and I spent a good deal of our early years. Across from the Beverly Center, it had an odd location for a rock club. Osco's had been closed for quite a few years. It reopened as the Cathouse in 1986. Taime and Riki had an apartment together and they were business partners. "The World-Famous Cathouse" was the place to hang out, and we always got treated really well there.

I remember me, Riki, and Duff would stair-dive into the wee hours. We would get all drunk, climb to the top of the stairs, and dive off, sliding down the railings and bouncing off the steps. Occasionally, I'd slice my ass or crack a rib. Great times, but God, how my body would ache in the morning!

## ⌐ NEW DIGS ⌐

The "splash sink apartment" didn't last and I found myself staying at yet another home, a little apartment on Martel, a few blocks from the Denny's on Sunset. This time I was sharing it with my friend Monica. She was a tall, gorgeous blonde, extremely fetching, who had moved from Sweden. She made about $200 in tips each night as a stripper at the Seventh Veil and went on to do some work in porn. I'd walk her to work in the evening around six and pick her up at around three in the morning. She'd put the cash in my hand, and I'd go nuts shopping for groceries at the twenty-four-hour Ralph's near La Brea on Sunset, known to everyone on the Strip as rock 'n' roll Ralph's.

A friend of mine, Cletus, had married Monica in order for her to get a green card. He was the drummer for a local band that played a few shows with us. Cletus and Monica, however, didn't actually have a real relationship. So he had no problem introducing me to her. There was an undeniable instant attraction between us. The very next day, I moved in with them. I slept in the bedroom with Monica, and Cletus had the couch in the living room. After about a month, it was as if Cletus was the one living with us. Monica and I would be camped out in the bedroom day and night.

Cletus and I would always play our latest demos for each other. He'd be like, "Listen to this," and I'd say, "Oh yeah? Well listen to this, fucker." We'd always find a way to challenge our arrangements, sharpen up the percussion, and it helped both of us. I'd go back in the studio and attack a tune we had been working on with a fresh fervor. I'd look up and the expression on Duff's face said it all. He was pleased. He knew I was working it. Duff and I were the keystone; we were the rock that rolled. We were to become the rhythm section for the biggest rock band in the world, and we pushed each other day and night to get there.

## ⚔ HELLO, OLD FRIEND ⚔

Meanwhile, Izzy had a new place behind Grauman's Chinese Theater in the heart of Hollywood. One afternoon, I popped by and walked in on him and Slash in the kitchen, where they were sitting down. Izzy had his eyes closed and his head back. Slash had a needle stuck in his arm.

My eyes bugged. "What the hell are you guys doing? That's sick!"

"Dude, it's dope," Slash said.

I looked at the needle they were using, and it made me ill. I *hated* the sight of needles. So I laughed to cover up my fear and just blurted out, "My grandma's got a whole box of those things." Big Lilly was a diabetic, and she always had syringes on hand.

Suddenly Izzy came to life. "What? Go get them. Go get them now!"

I shot over to Grandma's and came back with a handful of fresh needles for them to use. I swear, the look of those things going in their arms grossed me out so much, there was no *way* I was going to do that. Having forgotten my earlier ordeal at Bob Welch's house, Izzy set me up with some foil, popped a piece on it, and cooked it up. When the smoke curled off the smack, I smoked it.

Again I got so damn sick. I puked in their toilet for a half hour. As I gargled half a tube of toothpaste from the counter, I realized that although I wanted to hang out with my bandmates, this shit was definitely not worth the hell it put me through.

## ⇥ NAME YER POISON ⇤

Izzy and Slash could really handle dope though. Izzy was just so mellow and cool all the time, you never knew if he was on anything. And when Slash was on the shit, people just thought he was drunk. Duff didn't have a taste for the "brown," but he was a serious drinker and was always half-crocked. Weed was my thing. Of all of us, Axl seemed to be the most straitlaced. He'd drink and smoke, but I never saw him get out of control with any hard drugs.

Now, we all had a taste for coke, like at a party or something, but we always held it together. At that time, we never flaked on band-related events. The band was our responsibility, but we never talked about it as such because we were having too much fun. It was just understood that you didn't let your partying get in the way of what the band was about. You didn't let the band down.

The band's reputation as a standout kick-ass act became more widespread; word of mouth exploded and our popularity soared. We recorded some demos and handed them out to as many industry people as possible. KNAC, the popular L.A. hard rock/metal format, was the first radio station to play us. They had a show on Sundays that started at ten P.M. where they gave local bands exposure, and they played "Welcome to the Jungle" from the demo cassette. Slash and I were on our way to the Rainbow when we heard it on the radio. There is absolutely no way to explain the thrill of that experience. Hearing your song on the radio is one of those moments that gets seared on your soul. Slash just kind of chuckled while I went ballistic: "Yeah. Dude, that's us!"

As for the business end, it always seemed to take care of itself. Slash and I would hand out flyers day and night. We'd walk from the Hell House to the Strip, each taking one side of the street, and wallpaper that entire stretch with flyers. They were everywhere.

There was a print shop across from the Guitar Center where we would have them run off. Marc Canter and this Asian dude, Jack Lue, would take the photographs. Then Slash would take his artwork and create a flyer. Our stripper friends gave us money to have them made. Everything just seemed to take care of itself. We always had advertisements for our shows running in the free local L.A. club scene magazine *Bam*.

I had been reading *Bam* from cover to cover for years, but I never got over the thrill of seeing our picture in there. We were now delivering big crowds to our shows, and the local clubs knew they could count on us. I knew we were well on our way, and sure enough, in 1986, a seasoned pro who saw something in us offered to take us to the next level. We eagerly accepted her help.

icki Hamilton was a familiar face who was always on the lookout for new talent. It was no secret that we were becoming a major draw on the Strip, and Vicki was determined to capitalize on our popularity. Over the course of a couple weeks, she approached each one of us, either before or after our shows. She took the time to answer our questions and impressed us with the fact that she knew the business inside out and had no ego. I took an instant liking to her. She looked you right in the eye and didn't brag, blow smoke, or over-promise. She basically said her actions would do the talking and told us she had already booked us a show.

This was the first time that we didn't have to book a gig on our own. The general attitude among the guys was very simple and straightforward: as long as Vicki was helping us, hustling up something good for the band, she was a part of us. She continued booking shows for us and even gave us money from time to time. Vicki was from Indiana, so she, Axl, and Izzy kind of bonded. She had earned her stripes working with Mötley Crüe, Stryper, and Poison. I would have to say that out of all the guys, I was the most vocal about the fact that I was impressed with her. The other guys always played it closer to the chest with their thoughts and feelings. I appreciated the jump start she was giving our career. She really believed in us, and just that helped tremendously. I have to say that looking back, if it wasn't for her, who knows?

I have no idea why, but the five of us just up and moved into Vicki's one-bedroom apartment (we were broke, but I don't think that was the only reason). She shared it with another girl, Jennifer Perry, who became an industry person too, later working with Ozzy Osbourne. The girls took the bedroom, and we crammed into the living room with all our equipment. We had free rein in the place, and we would have chicks over and party all night. The phone rang nonstop, and there was something going on there 24/7. The apartment was in a small two-story building, and if I needed some privacy, I would just head up the fire escape to the top landing. Now, if that rooftop could talk . . .

## ⊰ BUSH BABY ⊱

remember I had this one girl up there who I had just met. She was a friend of a friend and had a nice hard body. We were getting it on, but when she lifted her top, I saw thick patches of underarm hair. Hell, you could have braided it and rappelled down the outside wall. I just cracked up, it was so funny. She was like this militant artsy chick who didn't go for all that shaving-everywhere nonsense. My God, she had the biggest bush. It was like being in the Congo. I needed a machete. She was great.

Whether at her apartment or at the clubs, Vicki worked her ass off for us. The first representative of a record company she brought in to see us was someone from Elektra Records. It didn't go well because we insisted on maintaining total artistic control over our music, and that was just unheard of at the time. But regardless, after word got out that Elektra had sat down with us, all the record companies became interested. Vicki set up the meetings with the record people and she would screen each one of them, knowing what we wanted. If she felt that a label was genuinely promising, then she would have us meet them.

## ⊰ AYATOLLAH AXL ⊱

She handled our press and got us a cover with a magazine called *Music Connection*. This was a widely circulated local music publication based in L.A. The interview was held at Vicki's place, and to show you how highly we thought of her, we insisted she be an integral part of it.

Izzy was so drunk, he kept interrupting everyone. Well, we all had been partying, maybe to try to calm our nerves about the interview. So we were all jumping in, just impulsively blurting out whatever came to mind. I remember that at one point, when we were talking about the way we created our songs, Axl said something like "I just want control over fucking everything."

So Slash jokingly compared him to Ayatollah Khomeini, who was not exactly a beloved figure in America. Axl got a little pissy over that. Then we all got pissy over the fact that the interviewer kind of jumped on the "total control" comment to see if he could get us arguing among ourselves. I guess the way Axl spoke, it could have been interpreted as he *alone* wanted total control and was not speaking on behalf of the band. Then we all kind of ganged up on this guy, because that's the way we were back then. You took on one of us, you better be prepared to take us all on. The next thing that happened was epic: Izzy shouted, "Fuck you and your magazine." You know what? They printed it. The writer ended his article by saying, "Well, fuck you and your band." That was great.

When we got the magazine about two weeks later, I was a little disappointed with the cover; I hated that picture, but we had no say over what photo they would run or copy they would print. I remember Axl was pissed because they spelled his name wrong: *Axel.*

There was a positive, and that was that the *Music Connection* thing generated even more word of mouth. Our shows were now selling out regularly and people who couldn't get in would just mill around outside. They sensed they were near something unique, something big. Vicki coming into our life had definitely moved us closer to our dream. Man, could she work the phones. She was very tough, a hard-ass at getting things our way. One time I heard her mention that she fancied herself a "white witch." Maybe she had read about Aleister Crowley, Robert Johnson, or Jimmy Page and actually dabbled in some kind of dark magic.

One night she introduced us to Tom Zutaut and Theresa Ensenat of Geffen Records. We could sense these people were the big guns by the way they conducted themselves. They took us to dinner. I think it was at Wolfgang Puck's on Sunset. It was very unusual for all of us to be in agreement but somehow this pair won over the entire band. After we were guaranteed absolute and complete creative control over our music and image, we knew that this was the way to go.

Tom was a very cool guy. He was all about giving us major freedom. It wasn't like "We'll only change this" or "Do it like this and you're in." That's why we liked him. Other labels pretended to go along with us but always tried to tack on some bullshit clause at the end. They wanted to control us and just make us some puppet band.

So we kind of knew we were going to go with Geffen early on, but—and this shows our playful mind-set at the time—there were still a few labels that hadn't taken us out to dinner yet. So we told Tom we needed a little time to think about it.

It might seem silly, but when you're flat broke all the time, getting free drinks is a big deal. We'd be at the table in a fancy restaurant and someone would yell for the waitress: "Cocktails." Then everyone would yell, "Cocktails!" Duff liked screwdrivers, Axl would get some fruity mixed drink, Slash liked vodka cranberries, and Izzy was strictly a wino. I liked Jägermeister, but I also liked beer or Jack and Coke, anything that would get you buzzed and tasted good.

## COCKS AND SNACKTAILS!

At one of the dinners with a record rep, we went to new lows. We were beyond drunk, joking around about who was getting the most head and some other lewd and rude topics. Evidently things got completely out of control, to the point that they wanted to boot us out of the place. I remember the singer from Chicago, Peter Cetera, was having dinner next to us, and he just stopped eating, looking completely disgusted. Somebody shouted our traditional demand for cocktails but it came out wrong: "Cocks and snacktails." We all burst out laughing. Then we compared cock sizes.

Eventually we got all the labels to wine and dine us: Sony, Elektra, and Warner. At one point, Megaforce was interested, and Rick Rubin wanted us too, but our minds were made up. We were just jerking these other record companies off in order to run up massive bar tabs.

# 10

# GETTING IT ALL DOWN

## ⇥ THE BIG DAY ⇤

On the night of March 24, 1986, Tom Zutaut came over to Vicki's to have a meeting with us. It was a beautiful evening, so we headed up to the roof. Tom went over his offer again, breaking every detail down for us, as simply (for our muddled minds) and clearly as possible. We pretended to give it some thought although we had already made our decision. We let Vicki give him the news that we would sign with Geffen the following day.

I usually woke up early, and the day we signed was no different. I was bouncing off the walls while the other guys were just waking up. They were much cooler about it. "Stevie, relax, calm down," they said. Oh yeah, how silly of me. I shot back: "We're only about to make our dreams come true." I guess I was always the kid in the band.

It was a sunny day, and everyone was together walking toward the Geffen building except Axl, who was nowhere to be found. We looked for him for over an hour and finally someone, probably Vicki, spotted Axl. He was on the roof of the Whisky! He was sitting in the lotus position, as if he was meditating. Classic Axl: "Look at me, look at me, watch me be different, watch me bust your balls by making us all late for the biggest moment in our lives."

Some photographer was walking with us as we made our way to the Geffen building, snapping away as we walked inside. We entered the main door, passed the secretary, who apparently was expecting us, and walked right up to Tom's floor. Tom and Theresa were there and on the desk in front of them the papers were already neatly laid out. We each had like ten things to sign. Vicki had a lawyer look over everything beforehand, so we had no worries. I had been waiting for this day all my life. We signed the papers and we each got an advance of $7,500. We went out and got drinks, had dinner, then everybody went five different ways and did their own thing, armed, for the first time, with more than a couple of bucks in our jeans.

Later we went to Guitar Center and bought equipment. We were offered wholesale deals on everything. I could have bought a bitchin' new set of drums for $1,200 bucks but I didn't really care to. After years of barely scratching by, I just couldn't shift gears like that and start blowing money. Besides, I had my own drum set and I was happy with it. I just added one more crash, the only piece of equipment I bought with my advance. Oh yeah, and I also bought a big bag of high-grade bud, then shoved the rest of my advance in my jeans.

## YOU CORK SOAKER!

Just after we got signed, we booked a show at Gazzarri's as the Fargin' Bastarges. We got that name from the movie *Johnny Dangerously* starring Michael Keaton. The bad guys in the movie always talked like that, mangling expressions: "You friggin' iceholes. You fargin' bastage! You cork soaker!" Even though we were booked under an alias, the show was packed. The timing was great because the club had been closed down for a while due to a riot there. We happened to play the night it reopened, May 31.

We were in the parking lot when we saw Kelly Nickels from L.A. Guns walking around, shuffling aimlessly like a kid who lost his mother. The band was going into the club from the back and I said, "Dude, what's happenin'?"

"Oh, I just came into town. I wanna see the show but it's sold out."

"Come in with me," I said, and he happily joined us. That night was an epic show. Armed with the Geffen contract, we knew we were on our way. So we bore down and played our songs with an intensity that went beyond what anyone was doing on the Strip at the time. Extended solos, long jams, and fucking loud—we were getting a reputation for being the loudest band ever (although the Who had made that immortal claim while we were still filling diapers, and a little later Slade took a swipe, a fucking great, *loud* band).

We were going to be huge and we never had to compromise. We did it all our way. We never had to sell our own tickets. We never sat around after shows to push our shirts or anything. That was the sort of stuff Poison was about, because they really were all about the business: buy our CD, buy our ball cap, buy our condoms. Not us. We just wanted to play music. We were so much cooler, and the kids knew it and responded.

Tom had the idea for us to go in the studio and record an EP under our own label, Uzi Suicide, which was actually financed by Geffen. The idea was pretty novel at the time, although everyone does it now. Our whole deal with Geffen was kept pretty hush-hush. Before he signed us Tom had even gone around telling all the A&R people he knew that he thought we sucked. But that's how Geffen operated, out of the box and pretty slippery. So they thought by making it seem that we financed a record on our own, it would contribute to our authenticity, the all-important street cred. As long as we could get our music out there, without anyone fucking with it, we went along.

Geffen wanted to put out the live album quickly and get people even more excited about us. It would also get us warmed up to record our full-length album. Honestly, we always had the idea to do a live record. Growing up our favorite records were live records: Kiss's *Alive!,* Judas Priest's *Unleashed in the East,* Cheap Trick's *At Budokan,* and the massive *Frampton Comes Alive!*

Frampton and I are a lot alike when it comes to performing. He's always smiling, always happy, working the crowd, reaching out to his fans. Look what *Comes Alive!* did for Frampton. He slaved for years with his band Frampton's Camel, putting out four studio albums with some incredible songs. But *Comes Alive!* put him out there and over the top. I think it's the biggest-selling double live album of all time.

Songs like "Do You Feel Like We Do?" and "It's a Plain Shame" were studio gems recorded by Peter like five years earlier. But when people heard them on *Comes Alive!* they flipped. Those tunes were made to be played live and loud. They were suddenly reborn and hugely popular. That's definitely one of the greatest live albums ever recorded. And if you listen to "Paradise" or "Jungle" on our later *Live Era,* you get the same rush, a realization: "So that's the way it's supposed to sound!"

Live albums transcend. They bring the full potential of a song to the audience. The way the crowd noise swells when Frampton slams into his first solo on "Something's Happening" gives you chills. To hear that same kind of intensity out of Frampton, you have to go back to the live album he did before *Comes Alive!,* and that was *Performance: Rockin' the Fillmore,* when he was still just a teenager playing with Steve Marriott and Humble Pie.

## ⊰ REALITY BITES ⊱

The idea was to have a "live" record with thousands of people screaming in the background, thereby making us sound as popular as, or maybe more popular than, we actually were. So yes, we knew from the start that they were going to add an audience. We were cool with it. Just so long as it sounded right. We didn't want this album to sound tinny or cheesy. Geffen's engineers told us there would be too much shit involved (i.e., it would cost too much) to actually record a live record, so we were told to create the live audience effects in the studio. Although I'll admit to being a little upset about the authenticity of it all, I ultimately felt it was okay because many of the live records we loved so much as kids weren't really live either.

They told me that was the case with *Comes Alive!* I was floored to find out that the only thing that was actually live on that album was the drums. Also, on Priest's *Unleashed*, Rob Halford actually recorded the vocals at Ringo Starr's house. I couldn't believe it. So we were learning the game and rolled with it, just so long as they kept their word and, as I said, didn't fuck with the songs. It was a bit of a tightrope for us, because we wanted to get our sound out there; we wanted them to know we were in it to win it, but we didn't want to completely bow to their direction.

Recording time was booked at Pasha Studios. Pasha was right next to Paramount Studios near Melrose Avenue in Hollywood. Spencer Proffer was hired to produce and it was his studio. Quiet Riot's *Metal Health* was recorded there, and that album was huge. We recorded the four songs, and I swear, I think we recorded "Shadow of Your Love" there too. Come to think of it, we may actually have done six songs during the session.

One thing always bugged me about the very beginning of the record. The count-in to "Reckless Life" is my very *first* hit on the drums. It's the high hat and cowbell. When I hit the cowbell the stick *slid off.* So my first recorded note is muted, it's not all there.

In the beginning of "Mama Kin" we added the sound of fire-crackers. If you listen closely, before the song starts, while Axl is saying, "This is a song about your fucking mother!" you can hear them going off: *crack-boom cracka-boom bam-boom!* We actually lit the firecrackers in the studio. We set them up in the recording booth, lit the fuse, and had them covered by a bucket. Of course the bucket was miked and it came out sounding huge. After we finished the songs, Spencer added the audience. He used archived tapes of live performances by Dio and Quiet Riot and mixed the cheers in. Spencer had been in the business a lot of years, and I really dug working with him. He had a lot of great stories, and I couldn't get enough of hearing them. He had done so much; I was very impressed. He worked with a lot of my idols, musical artists from the sixties and seventies. Plus he was a great human being and it was easy to work with him.

Every day around noon we would break for lunch and go to Astro Burger on Melrose, home of the best burgers in L.A. Then back to the studio, where the whole recording process took two or three days. We were all in the same soundproof room and we actually recorded those songs together to give it a "live" feel, instead of each performer laying down a separate track, then assembling the tune. The only stuff they overdubbed was the backing vocals. If you listen closely to "Nice Boys," you can hear Axl singing backup to his own vocals.

The record came out, KNAC put "Mama Kin" and "Reckless Life" in regular rotation, and it was an incredible thrill to hear my band on the radio. I experienced the most joyful, natural buzz from this. A movie that was released in 1989, *American Ninja 3,* featured "Move to the City" on its soundtrack, but I've never seen it.

We were at Vicki's when she came in with the first shipment of our record. It felt like Christmas morning. We just watched as she opened the box packed with EPs. It was about the size of the boxes that hold ten reams of paper in a stationery store. It was a feeling just like the one I had when Slash and I heard GNR on the radio for the first time. I experienced many fantastic firsts in my life at this time. The child in me couldn't get enough, as every morning smiled down with the promise of more and more artistic highs.

The cover featured a close-up shot of Axl and Duff. It was such a cool picture, the lighting, their expressions; I thought it was perfect. Everyone in my family bought a copy. Our very good friend Marc Canter bought a couple. The first store that I walked into and actually saw the record on display was Vinyl Fetish on Melrose. The owner, Joseph Brooks, was a close friend of the band and, like a lot of locals who had charted our rise, shared in our accomplishment.

Geffen assigned a personal manager for us, Alan Niven. He was a big, shit-talking tough guy with a British accent. He was also currently managing the established L.A. band Great White. I know the guys were hoping for Doug Taylor or Doc McGee to manage us, because they managed huge acts like Bon Jovi and Mötley Crüe. But Alan was raw and hungry, and he would be there for us. We all liked him. He was uncompromising and brutally driven, not unlike Zep's legendary über-manager Peter Grant, and he was gonna bust ass, get us busy, and get us to the top.

## ⊱ VICKI VACATES ⊰

All of a sudden, out of the blue, Vicki was no longer around. It just happened. At first I thought that she had cut some severance deal with Geffen and that was why she just dropped out of sight. I had heard no talk about tossing her aside when we got signed. I believed that she still had some tricks up her sleeve and would still have plenty to contribute to our success. I certainly got along best with Vicki; in fact, out of the entire band, I probably got along with all outsiders the best.

Slash really liked Vicki and Izzy liked her too. But I guess the band as a whole felt that she was not established enough, and in fact, a general feeling surfaced that a man would have more power. This was particularly true of Axl, who believed a woman would not get the same kind of respect as a man. Alan was a cool guy and never uttered a negative word about Vicki. This only confirmed our belief that he was going to be a consummate pro and kick major ass for us. I kind of made a mental note to find out the details of Vicki's departure, but in the swirl of getting the live record out, I never really followed up on it.

At this point, everything was happening so quickly. In the past, I'd felt that some of the gofers that we had around us were a bunch of desperate users who were out to leech off us and grab everything that they could. I believed Alan had successfully reamed out that grimy hole, and I felt much safer, less exposed to the greedy cling-ons.

We moved out of Vicki's place and set up in a roomy two-bedroom apartment right on the corner of La Brea and Fountain. It was, however, very rare that the five of us would ever be there at the same time. We were all over the place now, granting interviews, buying new clothes, checking out new equipment. Of the five of us, I probably hung out there the most.

## ⊰ KISS OFF ⊱

Paul Stanley of Kiss saw one of our shows and became very interested in producing us. He contacted Zutaut, and Tom arranged for us to meet with him. I was so stoked, I couldn't sleep. I never slept anyway, was last to bed and first up, but at least now I had a solid reason. Of all the surreally wild stuff that had been happening in the last month, this topped them all. I mean, we were about to be courted by rock royalty; this was Kiss, man!

Paul came to the apartment and sadly, almost immediately, the guys hated him. Paul probably knew as soon as he walked in the place that it wasn't going to work out. It just wasn't in the cards, and so he would not be producing us. The guys talked to him for about ten minutes.

Each guy would ask him something like, "Well, what do you think about such and such?" and Paul would answer with something that was probably the polar opposite of what we wanted to hear. One by one each member of the band just kind of drifted away. To be fair, I'm sure Paul felt he had to strut in with an authoritative manner to show us he could be in charge, but nothing, and I mean *nothing,* he said resonated with us. In fact it was more the opposite. I remember Izzy in particular didn't like Paul's response to one of his questions, and he gave a very shaky, "Ohhh . . . ," and then peeled off, saying softly, "See ya . . ." Within fifteen minutes the group was doing other things around the apartment, like jumping on the phone, digging in the fridge, watching TV, and not paying attention to Paul at all.

Eventually just my friend Ronnie Schneider and I were left. I was the last in the band to talk with him, and I was initially like, "Whoa. Paul Stanley." He was a hero to me.

But he wanted to change me, and that's where he lost me. First fucking words out of his mouth: "You need to get a huge drum set."

He told me this without explaining why. I just looked at him. "Well, fuck that," I thought.

I think we all felt that he wanted us to become the Paul Stanley Project. So I realized that I didn't want him to produce us, but I still wanted to talk to him. I was a big Kiss fan. I told Paul about my Kisstory experience and said, "When I was a kid, I would put my speakers on either side of my head, crank it up, and listen to you for hours." But by this point, he just wanted to leave.

I remained polite and walked him out. I think he wanted to get away from me because I was asking him all these goofy obsessive fan-type questions about Kiss. Then finally, at the elevator, I impulsively lifted up my shirt and said, "Who do you think has a hairier chest, me or you?" and he was like, "Well, *I* do, of course." He said it in such a snobby-ass way, I thought, "Oh well, you can have it."

We welcomed Paul, and I swear we all had open minds when he walked in, but I've never seen anything go south so quickly. It's because Paul came in with an attitude like, "You guys are the youngsters. I'm the rock star, and for this to work you gotta listen to me and do what I want." In the end, we weren't mean to him, we were just, "Whatever, dude."

## ⊰ AXL—ONE GREEDY MOTHERFUCKER ⊱

When the time came for us to record our LP, we moved in with Alan Niven at a much bigger house in Los Feliz. We began doing preproduction rehearsals at SIR Studios in Burbank. That's when the issue of crediting the songs, who got what, who owned what, and who got royalties for what, came up. It was Mike or Tom who told us, "You guys got to work this out. And you've got to have it all finalized before you start releasing your music."

So we gathered in the new place to sort everything out, just the five of us. Now, I thought it was kind of a formality because we had talked about all this before and from day one it was always supposed to be an equal share for everybody. But Axl had changed his tune. Axl wanted a bigger slice of the pie.

Axl didn't think it was fair to split royalties evenly five ways on our songs. He believed he was entitled to more than the rest of us. The other guys were smart. They just stared at the floor. No one said a fucking thing. I don't know if Axl intimidated them or if they just knew that silence was the best way to deal with his ego. Well, I couldn't just shut the fuck up about it. The reason I wouldn't dummy up was I was so outraged.

So right off the bat, I was like, "Screw you, I was here from the beginning, I worked on putting those songs together just as much as you." I had no trouble standing up to Axl because I was right. So now there's this deadly silence again, and it's obvious that it's become a big fucking deal. Still, no one else is saying anything, so rather than get into a big argument, I proposed what I thought was a fair offer: "Considering Axl *did* write most of the lyrics, which is a huge fucking part, I'll give you five percent of my twenty percent."

Axl shot me this look not of thanks, not of appreciation, but of arrogance and triumph. It was like he expected it. So I looked around the room because what I expected was for everyone else to follow suit and ante up too, but the room went dead quiet again. I looked around and everyone kind of started talking about other stuff. The matter was over, settled, done. Axl was happy and I was like, "Fuck!"

So it went 25 percent to Axl, 20 percent for each of the other guys, and 15 percent for me. The entire ordeal lasted only a couple of minutes. As long as Axl got more than everybody else he was a happy pig in shit. And at this point we were all trained to feel that as long as Axl wasn't being pissy, as long as Axl was content, then we should all be happy. He got away with more than the rest of us combined. Like climbing up on the roof of the Whisky the day we signed. If that was anyone else from the band, we would have climbed up there and thrown him off, but not our Axl.

We didn't know that Axl had a medical condition, manic depression, at the time. We just knew that dealing with Axl was tricky, that he was a moody motherfucker, and that you had to be prepared for craziness. One day he'd be hugging you and the next day kicking you in the balls. But Axl did some loving things for me that surpass anything the other guys ever did for me, so who am I to praise or condemn? I love the guy to this day, I honestly do. But that doesn't mean I'm going to lie to you about the way he was.

## ⇥ HIS OWN WORST ENEMY ⇤

Axl could get very uptight, while I was usually the opposite. People told me I was always easygoing. I got along with everyone and he didn't. Fact is, Axl had trouble getting along with himself. Axl was always living in his own little high-class snobby world, or at least he was in his twisted little mind.

I remember at this one show, he left after the first song because the monitors (the small speakers that face toward the musicians onstage so they can hear what they're playing) sucked. So he just split. As he stormed off the stage, he walked right by me. I shouted, "Why don't you come to sound check? Then you'd know what the monitors are going to sound like. You could even get it straightened out before the show." But no, that was asking too much.

Axl stood up thousands of fans without a second thought. One thing I've always respected is GNR fans, the most faithful, dedicated, fanatical audiences in the world. Unfortunately, Axl didn't feel this way, and after we became famous, he kind of took the GNR fans for granted.

Whether it was monitors or royalties, I was the only one in the band to call Axl out on his shit. Later that night we were in a bar and he's sitting away from the band with his latest bunch of "friends," who were lately shaping up to be B-list actors and wannabe models. He's shoving his smokes into a fancy cigarette holder, and he's looking fucking ridiculous. The other guys wanted me to leave it alone, but I couldn't, so I stood up and said, "Look at you, you pathetic little stuck-up motherfucker."

Axl just laughed at me: "Ha. Stevie, you're funny."

I go, "Motherfucker, what the fuck's wrong with you? You can't just leave us onstage and take off like that."

Axl just whispered something into the nearest ear, and all his sycophant friends tittered away.

When Axl was ridiculously late for a recording session or blew off an important gig, I felt I had to call him out on it. The other guys knew better than to draw the wrath of Axl, I guess. They would just look the other way and stow their feelings. But there were times when Axl treated me with twice the respect that anyone else in the band did, and I think it was because I was real with him. Somewhere in the depths of that tortured soul, he appreciated that. But eventually I would pay dearly for standing up to Axl, because I became the guy with the bull's-eye on his back.

Now, Izzy avoided hanging with crowds, preferring to be on his own. But he was respectful. He would go off with a woman and just chill out, surfacing when he was needed again. Duff, Slash, and I, well, we were always together. The three of us had a blast every time we went out. We were just born to party together.

## ⇥ RELOCATION BLUES ⇤

At this point, we started moving around so much, it really was just a blur. For a while we were staying in the house with Alan Niven in Los Feliz, by the observatory in Griffith Park. Duff's friends from Seattle and Axl's friends from Indiana ended up staying there for a while too. Then we moved to Manhattan Beach because Tom Zutaut lived there. He gave us a white van to commute in, and Slash was always designated driver. Of course, it wasn't long before our designated drunkard wrecked our ride. Good thing no one got hurt.

The time came for us to start recording at Rumbo Studios in Canoga Park. It was right next to the Winnetka Animal Hospital. It was close to my mom's house, and she cooked us lunch almost every day. One thing about Mom, she just couldn't stay mad at me for extended periods of time. I certainly took advantage of it, because I remember those meals came in handy. Mom brought us pasta, sandwiches, and salads, very tasty stuff. Then she'd ask if we needed anything and the guys would hint about running low on cigarettes, so she bought a few cartons for them. Then they went too far and gave her their laundry. And you know what? She even washed and ironed our clothes for us.

When we started working on *Appetite* we were in a hotel in Manhattan Beach, which was like a forty-five-minute drive to Rumbo. I have no idea why we were so far from the studio. One day my little brother came along with my mom to Rumbo. The band Heart happened to be recording their new album on the other side of the building. Their guitarist, Nancy Wilson, gorgeous and known the world over for her incredible songs, came by to say hi.

Nancy was very gracious. She lifted Jamie onto her lap and was very sweet to him. My little brother was smooth for a ten-year-old. He had the biggest smile on his face that day and soaked up every minute of it.

## ⊰ CLINK STINKS ⊱

Around this time our producer Mike Clink came up to me suggesting I change my drum setup. With all due respect, that's kind of like someone coming up to you with suggestions for changing your internal organs . . . you just don't fuck with what works. But I wanted to be a team player and when he got me a china cymbal and a second tom I was like, "Ah, what the hell," and reluctantly agreed. But the trouble with giving an inch is what happens next. They're not happy and they demand more. Maybe that's why it's better to be a miserable prick to people; they don't mess with you as much.

Mike asked me to change "Anything Goes" and that really hit a nerve.

"Fuck you, don't tell us how to write songs." I got so pissed because you don't meddle with the music. I pouted, stomped around, and behaved like a real dick. Where did this guy get off?

But I can't stay mad at people, and I couldn't in this case particularly since I knew in my heart that Mike was coming from a good place. So we tried his idea, and to my surprise, it came out great.

My resistance had just been from a deep-seated desire to guard our songs, and no one messes with GNR's tunes. But I will be the first to admit when I'm wrong or out of line, and after we worked it out, I looked Mike straight in the eye and said, "I am so sorry."

Mike's change happens right when Axl starts singing the first verse. It was initially at a slower time, and his idea made it faster, and like I said, better. So we started tweaking other things, like the chord changes at the end of "Rocket Queen." Also he had the idea to add a vintage Moog synthesizer to the beginning of "Paradise City" and again, that ended up sounding great. Those are the only changes I can recall that he made to the songs. At the time, "Mr. Brownstone," "It's So Easy," and "Sweet Child O' Mine" were our newest songs, and we worked our asses off on them in the studio.

"Mr. Brownstone" was a thinly veiled warning from Axl to all of us, including himself. We all saw how drugs had been granted a permanent VIP laminate in our lives, but we also believed we were indestructible. Although we were arrogant bastards, we respected (and feared) heroin's ability to weasel its way further into our lives, demanding increasingly bigger chunks of our daily routines.

So we did what we usually did with something that had become a part of us: we wrote about it. Same with the groupie scene, which was getting ridiculously out of control. We could just shove a fishing net out the window of any club and pull in choice catch after choice catch. The girl game lost its appeal; there was no longer a challenge to scoring the choicest snapper, and again, we chose to write about it: "It's So Easy." It was understood that Axl had final say over the lyrics, but we could all contribute, and at that point we all wanted to contribute.

# 11

# BUILDING *an* APPETITE

My contributions to the record took six days, start to finish, and I was done. On the other hand, Axl would insist on doing his vocals *one line at a time,* and that took much longer. Nobody wanted to be around when he was in the studio because his Talmudic recording methods drove everyone nuts. It was beyond what a perfectionist would demand. And it soon became obvious to us that it was obsession for the sake of obsession. Pretty soon the rest of the band just kind of slipped out to go to the bathroom and neglected to come back. We just preferred to be off campus drinking and partying while Axl was driving the engineers and techs out of their skulls.

While we were in the midst of recording, I remember Mike and Alan really didn't think the album was going to do shit. They felt that our songs were pretty much standard been-there, done-that hard rock fare. They were surprisingly vocal about it and it got back to us that Mike didn't think he was working on anything special. I remember a kind of ho-hum atmosphere in the booth at the time, and though I liked Mike, he certainly wasn't kissing our ass by any stretch of the imagination. There's a story about how when they shot our first music video, "Welcome to the Jungle," they felt the same way about it: nothing special. That might have changed because, as legend has it, when they were doing color correction of the final cut in the edit bays a couple of girls from the office slipped in to peek at it because they thought it was the "coolest video ever."

## ⚞ DEATH OF A FRIEND ⚟

After my tracks were done, I wasn't asked to get involved in any of the rest of the recording or mixing. Slash and Axl, however, went to New York to contribute to the final stages of the process. Joining them on the trip was our longtime friend Todd Crew. Todd had been part of the band's inner circle from the beginning. He was a shit-kicking, hard-drinking, exceptionally cool guy. He played bass in another band called Jetboy that originated in San Francisco. When they kicked Todd out of Jetboy, we were the first band to tell them, "Screw you, you're done as far as we're concerned. You're never gonna do shows with us."

Axl, Slash, and Todd flew to New York to oversee the mixing on *Appetite*. Todd never made it back. I don't know what happened exactly, because I wasn't there. I heard that he and Slash were partying, shooting heroin, and Todd passed out. Slash and Todd must have gotten separated at some point and Todd overdosed and died.

No one could believe it when we got the news back in L.A. It was the most terrible shock that I had ever absorbed at that point in my life, beyond devastating. I didn't want to eat, talk, or get out of bed. No way there could be justice in a world that would let a sweet beloved friend like Todd slip away. The band had friends who were so close, so devoted, that we considered them to be members of GNR who merely didn't appear onstage. Todd was one of these, and I truly felt I had lost a brother.

A week or so later, Slash and Axl returned. Their mood was beyond dark, and they avoided all calls. It was a horrible, bleak time for all of us, and it wouldn't go away. Each day there would be about a millisecond after I woke up where I'd smile at the sun; then it would hit me, and I'd spend the rest of the day reeling from my feelings over Todd's death.

Eventually, the dark clouds lifted because they had to. There was intense pressure from the label to get on with completing the album. If it were up to us, I think we all would have preferred to just lay low for a couple of months, but we were learning that many decisions were not entirely ours to make anymore. Massive amounts of money were pouring into the launch of the album, dates had been set, and commitments were carved in stone. But this is what I loved about my band at this stage in my life. Guns N' Roses was a living, breathing presence that knew how best to survive. No attitude, no petulance, just an organic desire to live and prosper. We didn't get rebellious or negative about the pressure; we just found a way to harness the love that Todd had shared with us and decided that rather than wallow in sorrow and self-pity, we could use *Appetite* to climb out of our depression. It worked, and I will tell you this: it worked because we honestly thought Todd would have wanted it that way, and no amount of bribery or bullshit from Geffen would have worked to make us finish *Appetite* if we hadn't believed that to be so. Death had knocked at the door, made us feel mortal for the first time, and GNR used *Appetite for Destruction* to rage against it.

## ⊰ APPETIZER FOR *APPETITE* ⊱

Slowly the work carried us through our painful recovery. To this day, when people come up to me and tell me that *Appetite* is the greatest record in creation and is the soundtrack for their lives, I believe that some of the magic in that album is owed to our love for Todd. The agony we had to work through pushed us further than any musicians had ever pushed themselves to deliver their absolute best onto that vinyl. And somehow we knew it; we knew we had cooked up something very special.

We got ahold of everybody who was anybody in our lives to get together at the Hell House for the "unofficial world premiere." It was to be our first listen to our new album. Wes, Del, the Naked Skydiver chicks, Jo Jo, everybody was there sitting around like kids waiting to see *The Wizard of Oz*.

Slash cued up *Appetite for Destruction* for the first time ever and as soon as "Welcome to the Jungle" came on, everybody cheered. Slash and I turned to each other and hugged; we were so happy. We listened to both sides, pretty much saying, "Oh yeah, that's working, that sounds cool," throughout. Everybody, *everybody*, was very impressed with what Axl did with his vocals. Funny thing though, I don't think Axl was even there. God knows what he was doing. Actually, I don't even think God knows what Axl is doing half the time.

"Paradise City" came on, and at the end of it, where it's got my drum fill that sounds like a double bass, I noticed something different. I know I did that fill only once in the studio. But Slash had the idea to repeat it somehow. I asked him right then and there, and he admitted the idea came to him in the studio. The second fill is actually the first fill played backward. There was a moment of tension as Slash looked at me like, "So I did it and it's done."

I smiled. "Dude, cool. Totally fucking cool." I had always played it, live onstage, with just the one fill. But it worked and it was completely all right with me because I respected Slash's call, and I knew in my heart that he did it to make the album as awesome as it could be, and that was my wish too.

## ⇥ FOR THE BAND ⇤

I guess I can justify my going along with Slash because I felt the fill worked and somehow improved the total climax to "Paradise City," or maybe it was because I was Steven, the smiling softie, the let's-all-just-get-along member of the band. Or maybe my self-esteem sucked and I wasn't willing to fight Slash in a battle I knew I'd already lost. Regardless, I didn't spend much time thinking about it, because the record was done, in the can. While we waited for its release, we found ourselves with a lot of free time.

Slash took a vacation in Hawaii. He had been partying really hard and needed to dry out. Physically, he was torn up pretty badly. His hands were trembling, shaking all the time. The trip was a personal get-it-together thing for him. There are some amazing party stories about us, but the truth was that we were all starting to show the wear and tear. It ain't the years, it's the mileage, and we had been on an incredible journey, working and partying harder than any dozen men. It was really starting to take its toll.

The fact is, when it came to drugs, everyone in the band was very private and secretive. Nobody told anybody anything about their preferences. Plus, we were all pretty greedy bastards. So when we were holding, we wanted it all. It wasn't like we were eager to turn one another on to our limited, or unlimited, supply. A few weeks later, Slash returned from the islands in considerably better health and spirits. We were a team again, ready to roll.

### ⊰ ALICE COOPER ⊱

In May, we were given a great opportunity to do a single show with Alice Cooper in Santa Barbara. We all were huge Alice fans, and it would have been our first really big show. Alan went out of his way to hook it up for us. The venue was a beautiful outdoor theater. The bill was officially booked, and we were so fucking excited. Imagine opening for one of your all-time favorite heroes. *Killer* is one of the greatest albums ever recorded: "Under My Wheels," "Be My Lover," "Desperado," "Halo of Flies," and the title track. I remember wearing out the grooves on both sides of that LP.

On the day of the show, we all piled into our new white van (we got another one after Slash totaled the first), while Axl was just standing there, outside. We were yelling to him, "C'mon, Axl."

He was all like, "Naw, I'll meet you there; some chick is gonna take me."

"Fuck that chick. C'mon, for the band."

That was our saying back then: "We're doing it for the band." That was our thing; we always said that. "We're gonna get drunk tonight—*for the band.*" Or "We're going to come on her face—*for the band.*" I don't know what Axl's issue was, but he insisted he get to the concert his own way, and what Axl wants . . .

Alice and his whole band were great. At the time Alice's band featured Kane Roberts on guitar, this bodybuilder dude who dressed like Stallone's Rambo. Eric Singer played drums. He went on to join my heroes in Kiss. Kip Winger was on bass, and he did a successful solo project a few years later.

They were giving us a fantastic opportunity, a great break, by letting us open the show. Hell, our record wasn't even out yet. We were ready to go, but sure enough, someone was still missing. Next thing you know, we're supposed to be on in five minutes and everyone is screaming, "Where's Axl?" We stalled as long as we could, but we really had to get out there out of respect for Alice.

At eight o'clock we hit the stage as scheduled. Without Axl, we just did our best and improvised. We did "It's So Easy" and Duff sang. After that, we just performed blues jams. We would always include a blazing blues jam in our sets, so we still managed to rock out for the audience, and I don't think they felt incredibly cheated. Izzy and Duff screamed a few words here and there. Duff's tech, Mike "McBob" Mayhue, may have sung something too. Bottom line was, without Axl present, we didn't deliver the true Guns N' Roses as promised. We just played, packed up our shit, and got out of there. Because of my worship for Alice, and my feeling about what Guns N' Roses was about, it was one of the most humiliating nights of my life.

Afterward, we were all pissed, and for one infuriating moment, we all considered kicking him out of the band. But we realized there was nothing we could do. The album had already been recorded and Axl was an integral part of our image and sound, so we never actually talked about getting another singer. I know this sounds like Axl got off easy, but we couldn't even dwell on it for that long. Shit was happening with the album release and we had to keep a calm head to make decisions.

Alan called us in for a meeting and he chose El Compadre, the Mexican restaurant across from Guitar Center on Sunset. He knew that by having it there, he would have the most luck in potentially bringing in all of us because we loved that place. Accompanying Alan was a white dude in his late twenties sporting an outrageous mullet cut. Alan said, "Boys, meet your new tour manager, Dougie Goldstein."

Dougie extended his hand to shake each of ours. He sported an infectious smile, ear to ear. He just seemed *so cool*. He was genuinely excited about working with us, and his enthusiasm was real. He told us that from that point on, we wouldn't have to worry about anything. He assured us that if we had any problem, anything at all, we could count on him. He had a winning way about him, a confidence, and we believed him. After the meeting, we decided to celebrate and hit up some other bars on the Strip. We all piled into Alan's truck. There wasn't much room for that many of us, so I said, "I'll sit in the back."

He responded angrily, "Fuck that, Steven, you're a member of this band. You're just as important as everybody else. Let me sit back there." I thought, "Now, that's cool. That's walkin' the walk." During the months that followed, I went on to believe I could trust him; I felt that I could tell him anything. I felt very close to Dougie, and in retrospect maybe *too* willing to share with him.

Later that month, Alan came to us and announced, "You all gotta get passports, we're going to England." Our *Live!? Like a Suicide* EP was hot, hot, hot and loved by both rock critics and our rapidly growing legions of fans. *Appetite,* our full-length record, wasn't out yet, so to promote our shows we prereleased a single, "It's So Easy," in the UK for our tour.

At last, I was off to see the world as I'd always dreamed. This was my first time out of the country, and I was bringing my rock band with me. We partied from the time we got on the plane until we passed out in our London hotel rooms. It was nonstop insanity. We were all pretty much travel virgins, except for Slash, who was actually born in England in a town called Stoke-on-Trent. I suggested we swing by the old homestead while we were over there, but Slash had no interest in visiting his birthplace.

When we arrived in England on June 19, it was cold and gloomy, and it remained like that for the duration of our stay. We were scheduled for three shows during the next ten days, and as I was a bit of a history buff, I became completely fascinated with the place. At night I would look at the narrow gaslit cobblestone alleys and think to myself, "Wow, Jack the Ripper probably stalked these very streets."

On the first day, we were taken in this sweet vintage-model limo to where we would be staying. We were put up in two little apartments, each featuring two bedrooms. They were living quarters for tourists who would come in and stay for a week or so. Just before we left, I had given Ronnie Schneider the opportunity to come along as my drum tech, and he had jumped at the chance, so he and I ended up sharing a room.

## ⊰ FATAL FLAW ⊱

As a drum tech, Ronnie didn't know shit. He was a bass player. But he was my buddy. He took care of business and we had a great time. There were a couple different tech guys in our crew: Slash had Andy, and Izzy had Scott, a guitar tech with really long curly hair, who was quite a nice guy. Mike "McBob" Mayhue was Duff's tech, and later he brought in his brother, Tom Mayhue, to tech for me. My attitude was "Hey, as long as I can get onstage and play, no worries." I didn't care about any of the other business pertaining to the band; as long as the basic shit was taken care of, I was happy. Later, this would come back to bite me on the ass big-time.

The place we stayed at turned out to be pretty damn old and sleazy (although coming from L.A., everything in England seemed ancient). There were cracks in the ceiling and walls, and crawling around were lots of weird-ass bugs the likes of which I had never seen before. So to get away from the Britannia bug safari, I went to a pub located across the street from the Marquee club, where we were scheduled to play. There I became friendly with the bartender, a Swedish chick. She was a cute thin girl who spoke broken English. I was chatting with her, and I complained about how there was practically no variety on TV in England and how I thought the BBC sucked. She told me that she stayed at the pub, in a room upstairs.

She had cartoons I could watch on her "video machine" if I'd like. Hell yeah. Did I mention this chick was hot? This was how it would happen for the band; everybody pretty much met their own people and carved out their own recreation when we were in London town or any town.

So I took cute little hot thin girl who spoke broken English up on her offer, and we hung out and got stoned. Later on, I discovered that she was a natural blonde and that she dyed her hair jet-black. I laughed and she asked why. I told her that in L.A. it was usually the other way around. We watched Bugs Bunny and I loved every minute of it. I spent most of the eight days we were there with her. We were in Amsterdam the other two days.

For rehearsals and our first show, we had to rent our gear. They gave me a white Sonar set, which sucked. We were only there for two days before the first show, but we rehearsed five times. Everyone was serious about establishing a solid beachhead: rule Britannia and you can rule the world.

On Thursday, June 11, we were ready to perform our very first gig in Europe. During the sound check, the guys started into a rocking song that I wasn't sure I had heard before. I was like, "Wow, this is a cool new tune." It had a haunting familiarity to it that I couldn't quite place. Since Axl wasn't there yet, Izzy and Duff started singing it the second time around and only then I realized it was "Knockin' on Heaven's Door." I smiled; "Oh yeah, it's *that* song." I realized we were taking the classic Bob Dylan tune and rocking out on it, taking it solidly under our wing into Guns N' Roses territory. That night we recorded it live, and it appeared as a B-side to the European release of our single "Welcome to the Jungle." We also rehearsed "Whole Lotta Rosie," the classic AC/DC tune, to perform at the upcoming shows.

It was Axl's idea to do "Knockin' on Heaven's Door." He told Slash about it, they learned it, and we did it. They never even mentioned it to me though, just expecting me to pick up the beat on the fly. I didn't know if this was a tribute to my drumming adaptability or a sign of their abject disregard for my needs as a member of the band (but I could venture a pretty good fucking guess). I should have put my foot down right then and there and insisted that I get as much time as everyone else to rehearse new tunes, whether they were well-known covers or not. I say this now because this growing disrespect only snowballed until it put me in an awfully embarrassing situation at Farm Aid.

We played the next Thursday, then again on Friday, June 19. The first show was great, although there were only about thirty people there. Afterward, I met a chick who, of course, wanted to party. I guess it should go without saying that I met a different girl after every show, 100 percent of the time. So I smoked a nice fat one with her. Usually after the shows, the band would go backstage, and we'd gather ourselves, get our heads together. I'd always take a shower, but I couldn't get one because the place didn't have the facilities. So this girl offered to let me take a bath, and she took me back to her apartment for a typical night of misadventure.

The next day during rehearsal, as I sat on my stool, I was experiencing some serious itching in my nethers. I was in so much pain. Then it dawned on me: I had crabs. "What the fuck? Dougie, help!" All the guys were cracking up, teasing me, but Doug ran right out and he got me this stuff, a shampoo called Rid. I went up on the roof of the building where we rehearsed, because it was the only place I could be alone and wash up. There was an outhouse-style washroom on the roof, a little room with a tiny bathroom inside. It had a hose to rinse off with. I filled the sink up with water. First I soaped up even though I was seriously freezing my ass off, then I rinsed the soap off with cold water. I scrubbed my entire body, lathering up a few times to be sure that I was completely rid of the critters. There I stood outdoors, stark naked, looking at all of England in the middle of the afternoon. Downstairs, the guys waited for me. Afterward I went back to the apartment, which had only a bathtub. It sucked. I *really* needed a shower.

The shows went really well. But for some reason, not all of the reviews were flattering. The press always seemed to want to "take the piss out of" GNR, and word got out that our first gig in England wasn't good when in fact it totally rocked! To be sure, there were some hecklers, but we won them over.

Now, *Kerrang!* magazine is the big weekly European rock rag. We did pics for it, and the photographer had the idea to have us lie all over one another. It became our first European magazine cover, out a couple of weeks before our show. We were doing so much stuff, photo shoots and interviews, twice a day, every day. I loved it, drank it right up.

Duff, Slash, and I walked around town, shopping at used-music stores and checking out the pubs, just like Mott the Hoople had done fifteen years earlier in the U.S. At some point during our visit, we took the ferry across to Amsterdam. While there, we received word that due to overwhelming demand, another show was added at the Marquee. We returned to perform a kick-ass set. The show went great, and we thanked the English fans for being so gracious.

On our final day, the limo came and picked us up at the motel. We were in great moods; we came, we conquered, and we'd be back. On the way to the airport, the sky became unusually bright. I looked out the window and thought, "Oh great, *now* the sun comes out."

Back in the U.S. we continued preparations for the release of the record. The choice for cover art was resolved pretty quickly. While checking out some shops on Melrose, Axl became quite taken by the artwork on a particular postcard that he found in a novelty shop. He purchased it and presented it to all of us. The title of the piece was *Appetite for Destruction* by Robert E. Williams.

We all liked it, and I was like, "Fuck yeah, that's perfect." It featured a monstrous demon with knives protruding from all its orifices and below it, a young female in distress, who presumably had just been raped. The people at Geffen agreed to run it and loved the opportunity to grab a little extra press, maybe a lot of extra press, with such controversial artwork.

Apparently, they were well aware that many retailers would never carry a cover that contained such a graphic misogynistic image. They went ahead and pressed the first shipment with it anyway, knowing that they'd probably have to change it for the future pressings. It was done intentionally to create resistance. Word got out about the offensive cover, and the story got some extra ink for the band. The black-and-white picture of the band on the inside cover was taken by Robert John at the Hell House. We were on the front porch with a keg of beer. The collage inside *Appetite* was designed to look like the inside of Aerosmith's *Live Bootleg*. Slash came to me with a stack of photographs and said, "Here, pick out seven or eight pictures that you want in the album."

The album was released on July 31 and the first shipment sold out immediately, because record buyers wanted an instant collector's item knowing that the controversial first pressing of the record would certainly be pulled. Slash and I went to Tower Records on Sunset and saw the little display they had made to promote the album. It was a bunch of our promotional posters and record flats glued together around words that said "$7.99 cassette, $11.99 record or CD." We just stared at it for like ten minutes. We were in some blissed-out primal state, so happy. We rejoiced: "We have an LP out that happens to be a kick-ass record. We did it!"

The day after the record's release, August 1, we shot our first video. It was described to us as a performance piece, which would be edited with graphic news footage and dramatized scenes of the band. The shots of us "acting" were first, and they created a couple of sets especially for the video. One was a big room with a bed and a TV, which was set up in an old dress shop. Another was a display of an electronics store, which they dressed up with TVs for sale in the windows. Then they placed Slash, half drunk, in front, drinking a forty in a paper bag, looking like a homeless derelict.

The following day we shot the live material. We invited all of our friends and filled the historic Park Plaza Hotel with an audience that truly loved us. We played "Welcome to the Jungle" live, five or six times, to get all the footage needed for the video. My little brother, Jamie, even made the cut. You can see him in the front of the crowd, pointing drumsticks at Axl. It did me good to see my brother Jamie in the video. There was someone who had seen me at my lowest but had only love and adoration for me. After we nailed down all the shots, we played a complete set for the crowd.

When we saw the finished video, it felt like another personal victory for me. Slowly but surely, we were building the appetite: we had the band, the label, the album, and now the video.

# 12
# TEARING IT UP *on the* ROAD

⊰ WHAT'S NEXT? ⊱

Ideas for tours began floating around. Originally the plan was to do a Midwest tour with Stryper, the Christian band I had dug on so much when I saw them play locally. They would throw Bibles out to the audience during their set, so me and Duff joked around about passing a couple of bottles out to the fans during ours. Another thought was to go up the East Coast with Y&T, another band that I saw often when I was younger. Those ideas, however, fell through. The very first tour to support our album was as the opening act for the Cult.

I remembered that a couple years before, Slash and I would dance at these clubs where they would play the Cult's videos on a big screen. In the early eighties, they called themselves the Southern Death Cult. They dressed all in black and wore white makeup, very gothic. Ian Astbury was the vocalist, a statuesque man with long black hair. He was so great. He would always let me go onstage, play tambourine, and sing in his mike. He just was the nicest, sweetest, most down-to-earth person. No wonder he had a beautiful, loving girlfriend.

You could talk to him, you could ask him anything, and he'd do anything for you. He made you feel good, and it was very comfortable to be around him. He was also a great performer. I didn't really hang out with their guitarist, Billy Duffy much. He seemed distant, maybe a bit egotistical. Les was the drummer and "Haggis" the bass player, whose nickname was inspired by the Scottish goat-stomach dish. Gross.

The Cult and GNR got along phenomenally well, and we had a great time together. They always had catering at sound check, great food that positively spoiled us. During our set, Axl made it a point to announce to the crowd how great the Cult was to us. As we would discover later, a lot of bands we opened up for would give us just half or even a quarter of the stage. The Cult was not like that at all; they gave us more room, more lights, more everything, the sign of a very cool, very confident group. When their album *Electric* came out, I really became a huge fan.

It's sort of a rock 'n' roll ritual for the headlining act to play a practical joke on the opening band on the last night of the tour. I was definitely the people person of the band, so I was always in with the roadies and the bands we toured with. The Cult's crew, and the band themselves, were all in on this particular joke. In New Orleans, during one of the last songs in our set, the Cult's crew came out and took my drum set apart piece by piece. First, the cymbal, then the cymbal stand, finally the snare drum, until I was just sitting there looking like a dork. Izzy, Duff, Axl, and Slash were all pointing and laughing at me. Then the guys brought the drum set back one piece at a time.

Now, usually opening bands dare not play a practical joke back at the headliners, but we got along so well, we knew it'd be cool. We got naked, with only towels wrapped around our waists. Then the five of us, and a couple of our roadies, walked out onstage while the Cult was playing. I had mixed a disgusting concoction of eggs, mustard, and relish in a Styrofoam cup. I walked behind Ian holding it. He didn't see me, and I motioned to the crowd, "Should I?" holding it over his head, ready to pour, and they were like, "Yeah!" He turned around and started chasing me all over the stage. He grabbed at me and pulled the towel off of my waist. I was totally naked onstage in front of everyone. I didn't mind one bit. In fact, I ended up without my clothes on many times, backstage, on the tour bus, in the hotels, and at the bars. The band called me "naked boy," a playful moniker and dependable indicator of how far along I was with my partying for the evening. I'd just look down and laugh—hey, I'm naked! Fortunately usually someone would wrap a tablecloth or something around me before I could get into any serious trouble.

That evening I covered my cock (needing both hands of course), smiled, and bounded off the stage. It was a thrill being stark naked in front of thousands of people.

The fun didn't end there. Afterward, I went upstairs to the dressing room, where Slash was talking to this hot little girl named Toy. He was looking to score with her but I walked in and she took one look at me and said, "Oh, I want to be with *him*." Thoroughly amused, I grabbed her, smiled, and said, "Sorry, Slash, that's how it goes."

Toy and I smoked a fatty and went out on the town. When we were leaving the theater, another hot young girl got my attention by grabbing my ass. She laughed and explained that she was a friend of a girl I knew in L.A. named Taylor. Taylor was a cool chick who had dated Axl and had been around the band from the beginning. This girl was from Baton Rouge and heard from Taylor that we were playing and came down. So I hit the town with a beautiful girl on each arm.

We went out on Bourbon Street. It was such a cool scene there. In one of the gift shops, I purchased a novelty cap that held a beer can, one on each side of your head. It had tubes attached so you could suck the beers dry. Wearing my new party hat, we entered a nightclub where we got drunk on Hurricane drinks. Toy had a couple hits of Ecstasy and since this was years before the drug caught on, I had never done it. It was mellow and pretty damn cool, a real body trip, like magic mushrooms. We were dancing, enjoying the lights and sounds like never before. All of us in the band had our own key to the bus, and the three of us went back to the bus and fucked and fucked and fucked. It felt incredible.

The tour dates for our shows started getting paced farther apart, and to pass the time we did what anyone who was bored shitless would do: we drank a lot. I also smoked a lot of weed while the other guys supplemented their booze intake with blow. Alan would pop up from time to time. And Dougie was with us all the time. He really made good on his pledge to take care of us. He had proved to be a real asset, particularly with his most important task: making sure that Axl got onstage on time.

Dougie ran everything. He was mom, wingman, and butler all rolled into one. He knew as long as Axl had hot tea with lemon, Izzy had his vino, and Duff and Slash had a steady supply of vodka, the boys were happy.

In late September 1987, we began a small tour of Europe again, this time with our good buds Faster Pussycat from L.A. The guys in FP were great. Of course, we knew Taime from years of partying at the Cathouse. Izzy and I really liked Brent, the band's guitarist. Me and Duff were hanging out with FP's drummer, a very nice guy, and after a night of hard drinking, he passed out in Duff's bed.

I couldn't understand it, but this made Duff super-pissed. Duff's the mellowest guy, but the booze could turn him into one mean mother. "Fuck this shit," he said. He wanted to play a practical joke on the guy, so he had me help him grab and tie the drummer's legs and wrists with duct tape. We taped all around his mouth and head too, and we carried him to the hotel elevator. It was one of those really old lifts with the gate that you have to pull open. We threw him in, and at that point, I thought it was funny as hell.

Then Duff pressed all the buttons in the elevator, closed the door, and let him go. The next day at the show, Duff and I saw him, bruised and very hungover. He avoided us completely, never uttering a word about the previous night.

When we got to Amsterdam, we went to the red-light district, where we met many stunning ladies. During that Euro tour, we hooked up with chicks everywhere. The girls were always there, always all over us. We were young, our dreams were coming true, and we reveled in it. Slash, Duff, and I would have contests to see who could get the most blow jobs in a single day. I won every time.

Slash and I would have orgies with five or more chicks. If I didn't like the way a chick looked, I'd send her over to the crew. Axl and Izzy weren't into that scene, however. They were more conservative; no orgies, no ménage à whatever, and I respected that.

## ⇥ AMSTERDAM GOOD TIME ⇤

Amsterdam is the greatest place I've ever been. Slash and Izzy were into heroin, and when they checked in, they couldn't wait to score some pure, quality shit. As soon as we got there, we all went our own ways in search of drugs.

For years, all I heard about from other rock musicians who had been to Amsterdam was how great the Bulldog was. The Bulldog's a popular bar in Amsterdam and I couldn't wait to hit it up. It was the first place I scoped out, and I was immediately directed to the Bulldog's pot bar downstairs. It was just a room full of thick sweet smoke.

On the wall were two menus; one had about fifteen different kinds of marijuana, and the other listed about nine different varieties of hashish. I was positively salivating over the prospects. Finally I said, "Give me the California Purple Indigo Bud" (I know, fly all the way from L.A. to Holland just to order some California bud). There were rolling papers on the bar in cups, much like napkin dispensers in a regular restaurant. They were huge cigar-size papers, Cheech and Chong's Big Bambú style all the way.

The Bulldog also had a drinking bar that was located upstairs. I stayed in the smoking area for the most part. It was amazing. Everybody's dancing, the lights are flashing, and I'm drinking beers and smoking bud, definitely feeling right at home.

The venue we performed at, the Paradiso, was located directly across the street from the Bulldog. At the end of the show, I walked to the front of the stage and said, "You rock my world. Thank you very much." I threw my sticks out to the people and jumped into the crowd. They gently set me down so I could walk right out the door back over to the Bulldog. That's how much I loved the place. I never wanted to leave.

Of course Slash and Izzy continued with their fix-ation. All they could talk about was scoring heroin. That was *all they talked about.* When they finally scored they were horrified to discover that the shit they got was fake. They got screwed and they were depressed, because heroin was supposed to be good and plentiful. The truth was that smack actually was there in abundance; they just had shit luck.

Ronnie and I were walking down the street, and we saw this derelict wandering aimlessly. Two police officers walked up to him, sat him down, searched him, and found used syringes on him. I just assumed he was busted. But no, they broke the needles, disposed of them, and handed him new ones. They also gave him a box that contained a syringe, a rubber, and an alcohol swab. Then they sent him on his way. I thought that was so great, so *enlightened.*

Later that night, Ronnie and I walked out of the Bulldog. We stood there for a bit enjoying the night, and then this guy came up to us and asked, "Hey, you wanna party?" I smiled. He said, "Well, a friend of mine has a little flat." He explained that he was a big fan of the band, so I figured he was totally cool.

We followed him into a dark alley behind a Holiday Inn. He told us to wait for him a minute, and he ran inside. We were standing in this alley for about twenty-five minutes and started getting impatient. Right when we were about to say "Fuck this" he finally came out and told us, "Okay, guys, it's cool, come on up." We entered the structure and it was pitch-black; I couldn't see a thing. We were walking up a long, spiral staircase. Someone had their ten-speed bike locked up in there, just where you couldn't see, and I tripped right the fuck over it. "Oh, watch out for the bike," he said. Yeah, thanks.

We entered into this cool little den, which was illuminated with red, yellow, and green lights. Heavy, flowing beads hung in the doorways, very retro-psychedelic. He introduced us to his friend Sven, a scruffy-looking fellow in his early thirties. Sure enough, he had heroin and coke laid out before him. I'd done heroin twice before, but I never shot it up. We were in Amsterdam, smoking buds, feelin' great, and we said, "Okay. Fuck it, let's party."

## ⚔ GO SPEED RACER ⚖

He had this brown powder heroin and a little pile of clean white cocaine. I asked the guy if he wanted some money for us to party, but he refused, saying we didn't have to pay. The guy reached underneath the couch he was sitting on and pulled out a spoon. To the right of him was a brown paper bag full of factory-fresh syringes. He took the spoon and dipped it into the pile of coke, and then repeated with the heroin. He was mixing up a speedball. I never did this; both times I smoked the shit I got so sick. I don't know why I was going to do it again, but I was just there, and that was reason enough.

He held the spoon over a candle and cooked it up. He dipped the point of the needle into the hot liquid and filled the syringe with the concoction. They wrapped a piece of cloth around my arm and tied it up nice and tight. I guess being in the presence of a pro lowered my fear of needles, because I just relaxed and stared at all of the colorful lights in the room.

He'd barely started plunging the syringe, and some red flag popped up in my head. I screamed: "Take it out, take it out, take it out!" I instinctively pulled away from him as he quickly removed the needle from my arm. I could see that I had gotten about a quarter of the intended dose. I was instantly in a dazed euphoric state, but I was barely able to hang on.

I swear, if he had shot the whole thing in, I would have been in dire straits. The rest of the evening he just poured us drinks and played great tunes. I settled down pretty quickly, and we went on to have an incredibly fun night.

Early the next morning we went back to the hotel. The sun was barely coming up and there's Slash and Izzy sitting inside, still completely bummed. I boasted, "Yeah, we partied all night. We were doing coke and heroin, we were fucked up." They were so pissed. "Why didn't you come get us? You asshole!"

That night, Ronnie and I went back to the Bulldog. After getting nicely lit, we left to go exploring and check out the local culture a little more. We walked down the red-light district, where hookers and sexual decadence abounded. Just like window-shopping, you could view all kinds of girls literally on display. They were dancing, gyrating, trying to sell themselves. Whatever you wanted, tall, short, black, white, stacked, you picked her out for an hour or half hour, whatever.

I didn't need that shit; I did just fine for myself. We passed by this one porn shop where this Middle Eastern guy waved us over. "I want to show you something." He pointed up to a TV monitor in a corner of the place and we were mesmerized. A man and a woman were fucking and sucking off every farm animal you could think of. We couldn't stop staring. We stood there for forty-five minutes, mouths wide open. The guy tried to sell us videotapes of the action we were watching, but no thanks; I could go a long time before seeing that woman with Mr. Ed again.

We finally decided to visit our local friends from the night before. We did the right thing and collected Slash and Izzy, then proceeded to the den of euphoria that had so rocked my world the prior night. After a small search, we found the site and knocked on the door. The same guy answered. He smiled and said that he had an idea we might be back. It became my second time ever shooting up. We were higher than high. This time they knew to give me a smaller dose, but Izzy and Slash said "fill 'er up" and it was speedballs for all as we raged all night.

After Holland we went to Germany to play Hamburg and Düsseldorf. All of our shows were practically sold out. The audiences loved us. The Germans were insanely into the show, singing along. They knew every word and it threw me for a loop. A lot of Germans spoke English perfectly. I remember hearing the band on the radio in Hamburg and jumping for joy.

The German cities were immaculate, like they had cleaning ladies come out and scrub them at dawn every morning. While in

Germany, I couldn't help but think about my family, the Jews, and the Holocaust. My grandma barely escaped to the United States just days ahead of the Nazi invasion of Poland. I shudder to think about all those innocent people packed into trains and shipped off to the gas chambers.

The next day we were to return to England, but I still had some weed from the Bulldog. On our ferry trip back, everyone was all worried and freaking out because I was bringing pot across and that was illegal. I wasn't concerned; I would just throw the shit overboard if I had to.

The weather was overcast, foggy, and cold. As we were crossing the English Channel, all I could think about was World War II and the invasion that was staged from those frigid waters. The whole turning point of the war took place there, massive maneuvers that brought tens of thousands of courageous troops from England into occupied France. Because of their bravery and sacrifice, I could play my music in a free world.

We continued our tour with stops in Newcastle, Nottingham, Bristol, and London. The shows started to meld together, but I distinctly remember the British crowds were a little more reserved than the Germans, although they clapped very enthusiastically.

Our last show in England was at the Hammersmith Odeon on October 8, 1987. We opened with "It's So Easy" and rocked that place, closing with "Sweet Child O' Mine" into "Whole Lotta Rosie." Playing there cemented our popularity, which had grown during the tour. The Odeon seated over thirty-five hundred, five thousand standing room only. From Cream to Van Halen, a lot of the biggest, most legendary bands played there, and it felt amazing when I looked out over the crowd. It was nearly packed, and those Brits caught one hell of a show. GNR was moving up over hallowed ground to the big time.

When we got back from Europe, we were scheduled to have some time off. It was our first break from the road in a long time. We had been touring nonstop since our show at the Whisky back in March. We arrived at LAX with a shuttle bus sitting there to pick us up. They dropped me off at the corner of Franklin and

Highland and just took off. I had nowhere to go and nothing to do.

Before this, they had put us up at the Franklin Plaza suites. A lot of bands stay there. I never thought it was anything special, so rather than check in there again, I got an apartment down the street from where I grew up with my grandparents on Hayworth. My new place was across the street from the elementary school where I met Slash for the first time. I went to see my grandma, and as always she was very happy to see me and this time, so proud. I wasn't in touch with Mom at this point, and it really wasn't for any particular reason. I just didn't think to call. I wasn't being mean or unloving; it just honestly never occurred to me to ring her up. Anyone who knows me should understand that it's not something I do to hurt someone deliberately. That's the way I am, and I've never really dwelled on it.

It really felt great to be back home. I looked forward to hitting the Strip, checking out the clubs, and catching up with everyone. It was time to kick back and enjoy L.A. again.

# 13

# HANGING *with the* CRÜE

Something about being back home made me want to reach out to my closest friends and just reconnect. I was really burned out on band business for the time being, and dialing up my two best friends and kicking back seemed like the perfect way to decompress from eight months of nonstop touring. So as soon as I got settled, I contacted my childhood chums Ricardo and Jackie.

I planned to meet them at the Rainbow. Driving over, I was all pumped up and realized how much I was looking forward to seeing them. Friends are the gold in my life; there is no currency more valuable. And while I will put up with a lot of shit from my friends, I never had to with these guys. They always treated me well, and they loved me long before it all took off. Those are the best, most precious buddies to keep for life.

As we were having drinks, I was delighted by the fact that these guys hadn't changed a bit. We laughed about that day in Jackie's backyard when Jackie and Ricardo wanted to be in their fathers' lines of work, and I said I wanted to be a rock star. On this day, our futures couldn't have looked brighter. The guys were so psyched about what was happening with GNR, and they were so excited for me. They went on and on about how great the album was and how much they loved the video. We realized that each of us had, in his own way, fulfilled his personal prophecy.

Ricardo told us a friend of his was staying at a fancy hotel up Topanga Canyon. He invited me to come over and party with them. Jackie had to be up early and passed on the idea but promised to stay in touch. Ricardo and I went to the hotel, and that's where I met Ricardo's pal Dennis. He was a very big, intimidating guy, but like a lot of guys with his build, he was just a mellow dude.

It turned out, conveniently enough, that he was also a coke dealer. He had hundreds of little square ziplock bags filled with rocks—not crack rocks, but coke pebbles, powder that's hardened. Each bag's contents weighed an eighth of an ounce, "eight balls." Dennis was very generous; he was like, "Go for it, dude, whatever you want."

"All right!" I was truly appreciative. I remembered smoking the shit at Bob Welch's house, so I decided to get some baking soda. Problem was, I never actually cooked it up while I was staying at Bob's. I watched Ted take a mix of coke and baking soda and pour it in a glass tube. The tube he used was actually a cheap glass cigar container. He would then pour a little water in. He would heat the tube, waving it over a stove, and in a few seconds, *clink,* a hard piece of crack had formed. But I just couldn't do it. I would end up with mush. I'd add too much water or baking soda and never got the hang of it.

I must have wasted $3,000 worth of coke trying to get that little "Kernel Clink" sound. Thankfully, I didn't pay for it. Dennis was impressed enough with me, the *rock star,* that he didn't seem to mind at all. I remember being in there totally wasting all that coke. He had so many bags and I just kept trying and trying. Ricardo was a very responsible guy and even though he was partying with us, he still went to work in the morning. Dennis and I hung for a few days before I had to get back to business with the band. After that, I would hook Jackie, Ricardo, and Dennis up with passes and tickets whenever we were in town.

## ⊰ BACK ON THE BEAT ⊱

After checking out of Topanga, I went to a sushi nightclub in Hollywood. There, pop artist George Michael was waiting in line. I got in line too, and George took an instant interest. He was talking with some friends, but his gaze never strayed far from me. Once inside, I noticed him again, looking my way, trying to flirt with me. He sent one of his boys over. "*George Michael* would like to buy you a drink." This was before George publicly came out of the closet. His sexual orientation, however, wasn't necessarily in question. I graciously declined.

In the band, no one knew where the other guys were, nor did we care. Doug really kept it together. He again got us rooms at Franklin Plaza so we would all be reunited in time for our next venture. Doug would make up an itinerary every day. He would post it on the door of our hotel room or he would slide it underneath. It outlined what time to get up, what time to be on the bus, what time we were leaving, plus scheduled interviews and such.

On October 16, 1987, we began a new leg of the tour in Bay Shore, New York. There was a Japanese metal/rock band, also signed with Geffen, that was opening the shows for us. They were named EZO. They wore crazy, bright makeup, like Kiss, and their U.S. debut album was produced by Gene Simmons. They spoke broken English but they had an American crew. The band themselves did not associate much with us, but when they did, they were terrific.

## ⊰ TRUE LOVE ⊱

We played in Baltimore, Maryland, on October 18, and then it was on to Philadelphia, Pennsylvania. This was where I met my first true love. We were playing a club called the Trocadero. I went to sound check in the afternoon. There, Fred Coury, a genuinely nice guy from the band Cinderella, introduced himself to me. We were hanging out, walking around, shooting the shit, and smoking cigarettes. We entered the Troc from its main entrance. We opened the big doors, and there, standing at the bar straight ahead, was the finest ass I had ever seen. It was so tight you could palm it with one hand. She was standing with her girlfriend, and when she turned to talk with her, I saw just how perfect the whole package was. I said to Freddie, "I gotta meet that girl." Freddie said he knew her. "Let's go, right now."

We walked over and Freddie introduced us. Her name was Cheryl. I was already instantly attracted to her, and after just a few minutes of conversation, I knew that this was the girl for me. She was so down-to-earth and so sweet. We hung out that entire after-noon, just getting to know each other.

That night we played our show, and after, I brought her back to my hotel room. We talked most of the time and kissed a little, but I was so impressed with her, I really wanted to take this slow. I chose to treat her right. I got her number, and I would call her every night thereafter. We talked about everything and I discovered that she'd had only two boyfriends. She told me she had sex with each only once, and each time lasted about two seconds. I sensed that she wasn't like so many other girls who were all "take, take, take." She wasn't look-ing around to see if something better might be coming along. She hadn't fucked a bunch of guys and she wasn't a user. Shortly after, I would fly her out to be with me in various towns and cities.

Surprise! I was deeply in love. More on this a bit later.

Even though we had shot it, our video was not getting played. David Geffen had to call in a huge favor from the head of MTV to get one fucking airing of "Welcome to the Jungle." They tried to bury it at like five A.M. on a Sunday morning. But guess who's wide awake at that hour on a Sunday and just getting in from a night of partying? That's right, kiddies, GNR Nation! Legend has it that "Welcome to the Jungle" hadn't even gotten done with its one airing and the MTV switchboard was lighting up like a Christmas tree. They were all demanding to know one thing: when would MTV be airing the video again?

Soon we started getting played in regular rotation, and our popularity grew and grew. We went from playing clubs to theaters very quickly. On the big tours, we were being told that more people were coming to see us than to see the headliners, like the Cult and later, Iron Maiden.

When we were on tour, we would take care of business, and it would be time to hit the road for the next city. Often it would be around three A.M., and there would be no sign of Axl. The parking lot would be completely empty, except for one car. And there, inside, would be Axl and whatever groupie he picked up that night. We'd yell, "Axl. Come on, fuck her already. We're going." He'd yell back, "Fuck you!"

## ⊰ DUMP THE BITCH ⊱

Finally, reluctantly, he would join us, insisting that the object of this night's desire come along. This eventually caused a lot of commotion, because shortly after we'd leave, Axl would find something fatally wrong with his date and turn on the poor girl. We would sit there in stunned silence as Axl would make a big show of getting rid of her. We knew better than to get involved.

## ⊰ NEW YORK CITY ⊱

After we played the Ritz in New York, we were invited to appear on *Headbangers Ball*. It was very exciting. We went to the MTV building, and everyone treated us like the celebrities we were rapidly becoming. "Can we get you anything? Would you like a makeup or hair person? Have you eaten?"

They had people literally running around, catering to our needs, and I guess it was cool. You'd look out the window and see traffic and the people walking around Times Square. They had gold records and promotional posters all over the walls.

Interviewing us was a guy named Smash, the host of the *Ball*.

**Parental Units. Mel, Mom, and me.**

**Keeping the Faith.**
"Stormin' Norman"
and "Big Lilly" are very
proud grandparents
at my bar mitzvah in
1978. *(Deanna Adler)*

**Adlers All.** A
rare Adler family
reunion on the
occasion of
Jamie's high
school graduation,
1993. *Clockwise
from left:* Kenny,
Deanna, Mel, me,
and Jamie.
*(David Sears)*

**Big Lilly's Boys.** Kenny and me with our grandma.

**Young Gun.** Me at five years old.

**Leather and Lace.** I've cleaned up for a formal photo session with Jack Lue, October 1985. *(Jack Lue)*

**Ass Backward.** Slash and I try to get Axl pointed in the right direction to take the stage for our first sold-out gig at the Troubadour, 1985. *(Marc Canter)*

**Gourmand Goofing.** Guns N' Rotelle during a fun Jack Lue photo session in 1985. (*Jack Lue*)

**Stand and Deliver.** GNR rocks out on January 4, 1986. At this show, "My Michelle" was played live for the first time. *(Marc Canter)*

**Adler's Attitude.** Always happiest behind the drums, I'm debuting "Nightrain" live at the Music Machine, December 20, 1985. *(Marc Canter)*

**Welcome to the Cathouse.** Guns N' Roses loved playing "the pussy palace." This was a special video shoot for "It's So Easy," October 10, 1989. *(Marc Canter)*

**Guns N' Ronnie.** GNR gathers backstage at Fenders Ballroom five days before we are signed. Joining us is Ron Schneider, a great friend and musician, who was also my tech assistant. *(Marc Canter)*

**Giant Step.** Backstage with Duff before opening for Aerosmith and Deep Purple at Giants Stadium, August 16, 1988. *(Marc Canter)*

**Smack Down.** Slash and I wrestle at friend Marc Canter's house while Jack Lue referees, 1986. *(Marc Canter)*

**Slash Trashed.** Slash and me on the road. *(David Plastik)*

**Big Buddy.** Kevin DuBrow and me.

**Fast Friends.** Howard and I hit it off the first time I was on his show. A stretch limo, porn stars, and weed helped prep me for my appearance. *(Jamie Adler)*

**My Idol.** Frankie Valli of the Four Seasons with me.

**Wild Things.** Steven Tyler, Slash, and me during the shooting of Sam Kinison's "Wild Thing" music video.

**Drummer Boys.** Fred Coury sits in for me (note the broken hand). *(David Plastik)*

**Pug in My Pants.** Me with my dog Shadow.

**Meatloaf Concert.** Caro and I snuggle before a
concert outside Las Vegas.

**Appetite for Confection.** My wife, Carolina, and I gather at a favorite lunch spot in the Valley with friends Steve Sprite (*far right*) and Lawrence Spagnola. (*Morgan Saint John*)

**Alice N' Adler.** Opening for Alice Cooper, October 23, 1986. This was the night Axl got held up and we played without him. *(Marc Canter)*

This, however, would be his last appearance, as they were completely revamping the show. We taped the show on a Wednesday to be aired on Saturday, but we weren't down with how it all worked. Slash was reading off some tour dates and he announced the show at the Ritz in New York for Friday night, and Smash corrected him by saying it already went down the night before. Slash was like, "No, man, it's this Friday," totally forgetting the airdate of our interview.

Since it was the last show of the "old" *Headbangers Ball*, Smash asked us to trash the set at the end of our piece, which we did gladly. Before going to a commercial break, Smash said, "C'mon, boys, on three, let's rock. One, two, three . . . let's rock!" Nobody in the band knew what the fuck he was talking about so we were totally silent. He must have thought this was some group conspiracy to make him look bad. "Well, thanks, guys," he said sarcastically.

Next up was a show at CBGB, the famous punk rock club in Manhattan. Duff was particularly excited because his heroes Iggy Pop and the Ramones had played there. A lot of my favorites like Blondie and Talking Heads had started out there too. When we got there, I said, "Are you sure this is CBGB?" It was the smallest room, very, *very* intimate. It held only like fifty to seventy people. I just couldn't imagine that all those famous bands had played there.

We performed an acoustic set and I rocked the tambourine. We debuted some songs that we hadn't played publicly yet. The lyrics "I used to love her . . . but I had to kill her" from "Used to Love Her" got a huge laugh. And "Patience" got a very nice response.

We also played "Mr. Brownstone" and "Move to the City." Someone yelled out, "Drum solo!" so I shook the tambourine wildly. Everyone laughed. After CBGB, we played the Horizon in New York on Halloween. Then we went to Washington, D.C., and after the show we went out with the crew—*Mötley Crüe*.

Previously, in the winter of 1986, our publicist at Geffen, Bryn Bridenthal, had invited us to the record release party for the Crüe's *Girls, Girls, Girls* album. It was at the Strip Club on Sunset, the same place where they later shot the video for "Girls, Girls, Girls."

We were socializing and partying while listening to the new record. I thought it rocked. We talked with their people, who loved *our* record and thought *we* were a great band. It was so amazingly humbling to me. These guys were my heroes.

In Washington, we met vocalist Vince Neil, drummer Tommy Lee, and bassist/songwriter Nikki Sixx. We didn't meet guitarist Mick Mars until we were touring with them. Mostly it was Tommy and Nikki hanging out with us. We got along so famously we pretty much knew from there on that we were going to do a tour sometime. We just *had* to.

## ⊰ THE SIXX SHOOTER ⊱

That Christmas Eve, Nikki invited me to hang out over at his pad. He had a bunch of girls over. We were drinking and partying it up. Nikki asked me if I wanted to do some coke. "Hell yeah!" We went up to Nikki's bedroom, where he had this huge walk-in closet. We went in the closet for some extra privacy. This was where he had his paraphernalia. He retrieved a tablespoon and a syringe from a hidden compartment. He mixed the coke with a little water in the spoon and sucked it up in the needle. He shot me up because I didn't really know how to do it. The feeling was great, not what I expected. I wasn't freaking out, like all anxious or something. I saw this leather jacket hanging there in the closet and I said, "Dude, that's a great leather jacket."

"It's yours," he said. It fit me perfectly, and that made me feel a little more entitled to his amazing gift because, well, I knew it couldn't have fit him and his six-feet-three frame.

After shooting up coke, we continued with our orgy. There were so many hot young girls there, finer than fine. They were the hottest chicks I had been with at that point. The oldest couldn't have been twenty. They were wearing lingerie and silky nighties. They were making out with each other and eating each other out. They had a dildo and were fucking each other with it. I was aroused the whole evening and blew at least three loads. At one point, Nikki and I were sitting on the couch getting blow jobs. We had seven or eight girls, with at least three girls sucking our dicks at any one time. It was great. It was beyond great. Think of partying with all that prime flesh. Okay? Now dream of how it would be to do it at a party with one of your all-time rock idols.

Nikki really knows how to work it with the ladies. When we were ready to come, they all crammed in really tight together in front of us. They put their heads together with their mouths wide open, anxiously awaiting our climax. We both shot our come on their faces. I yelled, "Make it sick!" and they swapped our come through tongue kisses, licking it off of one another. It was awesome.

Nikki is a smart, all-around together, down-to-earth, professional, cool guy. GNR originally wanted the Crüe's team, Doc McGee and Doug Taylor, to manage us. Mötley had heard all about us from the club days and they had read about us in magazines. They came to our shows, they dug us, and we started hanging out. When November 1987 came around and the opportunity arose, we told Alan Niven, "Dude, we've got to do this tour with Mötley Crüe." Shortly thereafter, it happened.

The pairing of our two bands electrified fans of both camps. Each band was in its kick-ass prime and I think we both delivered the best rock 'n' roll show on earth in a long time. The first show was down south in Alabama. Late that night, after an amazing performance, Tommy invited me into the hospitality room where people would wait to meet the bands. They had catered food set up on three big six-foot-long tables.

Tommy put his arm over my shoulder and said, "Stevie, come here, I wanna show you something." He brings me into the room and closes the door and says, "Blow your nose real well." I looked down, and on one of the tables were two lines of coke spread the entire length, six feet long. I smiled and yelled, "All right!" He handed me a short straw; he started on one end and I started on the other. *Snort!* We met in the middle, just looked at each other, and laughed. He fell back on a couch behind him, and I fell back on another couch, and we just sat there for at least ten minutes. Well, it could have been an hour. Who knows? We finally got up, wired to the max, and finished the rest of the "krel." That's what they called coke back then. "Got any krel? Got any krel? *Où est la* krel? Here comes the krelly, man!" All the guys in Crüe were great. Vince seemed kind of involved in his own coolness, though not nearly as stuck-up as Axl was becoming. They both shared a sort of "too good for you" attitude. Mick Mars was very quiet and shy. I got to know him a little better on their private jet. We were thirty thousand feet in the air, and Mick made me a drink. It was the first martini I ever had. It was awesome, and it really gave me a taste for the art of a nice dry martini. The plane was the way to tour. We made it to the next town in like forty-five minutes.

At the time, Tommy Lee was married to Heather Locklear. She was the hottest thing on wheels back then. Whenever she would come out to visit Tommy during the tour, everything was "hush-hush." We couldn't talk about girls or drugs. We'd have to stop partying and be on our absolute best behavior. Truth be told, it sucked when she was there. Tommy had to act like a saint, although the night before we'd be getting blow jobs backstage from a dozen groupies.

The Mötley tour lasted only a month, and on the last night of the tour we were in Florida. Tommy had a drum set that was built in a cage. It would rise, go out twenty feet above the audience, and rotate vertically 180 degrees with Tommy in it! I said, "Dude, Tommy, you gotta let me try that." Since it was the last show and I was buddies with Tommy's crew, he gave the go-ahead.

After sound check, they said, "Dude, you wanna do this thing? Let's do it now." They strapped me in, the set rose, and they flipped it around 180 degrees and just let me hang there upside down. I was trying to play while all this was going on, but I had to lock my feet into the base of the snare drum stand to keep myself anchored properly. I couldn't figure out how the hell Tommy managed to keep playing. So I'm hanging and they're all, "See ya, Stevie." They start walking away. "Guys? Guys? Okay! Blood's rushing to my head. Not good!" Finally they came back, laughing their asses off.

Now, remember, on the last night, it's traditional for the headliners to play some sort of joke on the opening act. This show would be no exception. We were performing our last number, "Paradise City," when all of a sudden what appeared to be cocaine came pouring from the rafters, snowing all over the stage. It wasn't actually coke, but rather flour, massive amounts of it in the air. It was so damn funny. Anyway, sweat and flour do not mix. I was washing dough out of my hair for two weeks.

We got along so well with the Crüe that we felt it would be okay to get them back. We did the same prank we had played on the Cult. During the Crüe's set we set up the gag. I had the cup containing a gross mixture of eggs, relish, mustard, and mayo. I walked up the drum riser, stood by Tommy, faced the crowd, and held the cup up for their approval. They cheered me on. I looked at Tommy, said, "Hey, buddy," and slammed it down right on top of his head. All the gooey ingredients splattered over his face and hair. He was definitely cranky about it. While he was playing, it got in his eyes. I could tell that he was irritated, so I wiped his face for him. After the show he stared me down, shaking his head. "You fucker, man." But we were good.

## ⇥ ONE DUMB FUCK ⇤

That night, Nikki, Tommy, Ronnie, and I were in the motel room doing coke and Nikki suggested that we cook it up. I knew this would take a while so I cut out to get more ice for our drinks, figuring I'd be back long before Nikki was done with the alchemy. Coming back though, I couldn't fucking remember which room we were in! I knocked on every door on that floor, but nobody answered. I ended up roaming the halls in total agony, realizing I wouldn't be partying with them on our last night together. It was such a bummer. Come to think of it, the way they were probably in a frenzy cooking up and smoking the shit, there was no way they were going to open the door anyway.

When the Crüe tour wrapped, we jumped on the Alice Cooper/Ace Frehley tour in early December 1987. I had become friendly with Tommy Lee's "krel" dealer, and he told me, "Dude, I'll take you to the airport, we'll get you some stylin' new tennis shoes and a pile of coke." So I scored some coke and a sweet pair of British Knights. I hid the blow in my carry-on while he drove me to the airport. I got on the plane and flew to Dallas, where we would open our first show with Alice. As I got to the curb to hail a cab to the show, I suddenly started to feel very tense about running so late. I jumped in the cab totally freaking out and said to the cab driver, "Dude, take me to where the concerts are. You got to get me there right away." I just had blind faith that the cabbie would know the place I was talking about. I had like fifteen minutes to get to the place.

The hack drove like a wild man and I got there two minutes before we were scheduled to go on. The guys gave me mixed looks. Some were pissed and others just worried. Dougie had my stage pants and a fresh shirt all ready for me to slip into, and I ran right out to the stage with the rest of the band.

Afterward, I explained that the cab driver didn't know where the fuck he was going. But the whole experience really shook me up and I swore to myself I'd never get separated from the band like that again.

When we got up to Madison, Wisconsin, a couple of weeks later, I finally had a chance to talk with Alice. I had resolved to have this conversation because it had bothered me for a long time. "Hey, Alice, remember that time when you let us open for you in California, and we kinda blew it?"

Of course he remembered. "Yeah?"

I told him I wanted to apologize. I told him how much I worshipped his music.

Alice didn't blink and said, "Don't worry about it." Alice is the best.

Sadly, Alice's father was mortally ill. Initially, they were telling us that the tour was canceled. And we were bummed. Then a few minutes later, they informed us that the tour was back on. Then word got to us that no, we were packing up and going home. We couldn't get a definitive answer out of anyone. Then it became ridiculous. We're going home, we're not, we are, we're not. I got so frustrated listening to all these clueless fucks telling us what was going on when they really had no idea. I said, "Fuck it. I'm going to the bar, let me know when you figure it out."

This precipitated the beginning of a series of pretty self-destructive events that eroded my status with Guns N' Roses while intensifying my occasional bouts of low self-esteem. In fact, almost missing the show in Dallas was nothing compared to what happened during the next few months. All these little things began to add up, although I wasn't really aware of it at the time. That afternoon, I went straight to the local watering hole Slash and I had discovered the night before. I was so depressed; I slammed something like twenty kamikaze shots. I got terminally shit-faced and became pretty damn obnoxious. I don't remember exactly what I did, but the bouncer pounced on me, grabbing my legs, and another guy got my arms, and they threw me right out the front door. I remember bouncing up real quick, screaming, "Fuck you guys!" I charged the front door, but they slammed it shut, so I punched the metal casing that covered the light on the bar's front door. I wanted to smash it but broke the little finger on my right hand instead.

# 14

# EVERYBODY OD TONIGHT!

### ⊰ DOWNHILL SLIDE ⊱

My finger swelled up like a sausage, but I was too drunk to feel a thing. I staggered back to the hotel and entered the lobby screaming for Dougie. At some point, "naked boy" had reappeared, and hotel guests were flipping out. Fortunately Dougie was at the front desk, checking out or something. He spotted me, took one look at my finger, and his eyes bugged. My entire hand had blown up, making it look more like a foot. He said, "We gotta get you to the hospital." I spun away and ran out of the hotel into the parking lot. "No way!" Dougie had to chase me down as I darted between the parked cars. We attracted a nice little crowd before he was able to drag me to the hospital in my birthday suit.

After that incident, things started to accelerate downhill. The band was just like, "What a dumbass, breaking his hand." They didn't care about me one bit. No one called the hospital while I was there. No one gave a shit. There was no talk of postponing anything until I knitted up. They just went out and got someone else to fill in.

I swear, if it was *anybody else* in the band, they would never have gotten a replacement. No way in hell.

Now, maybe this was the blow talking, maybe the painkillers for my hand were messing with me, or maybe it was the recurring shitty self-esteem, but I began to harbor this growing dread that Duff and Slash didn't think I could play the drums that well. It wasn't anything they said; it was just their general attitude toward me at the time. I could tell they didn't think I was a good drummer, and I started to think they didn't think I was so cool either. Breaking my hand on a bar light didn't help. But I was just feeling very down at the time. Taking myself out of commission and off that drummer stool, however temporarily, was like taking away my identity.

Fortunately Dougie was my go-to buddy and he was very good at propping up my spirits. He had me believing everything was fine and I could count on him to have my back.

We said good-bye to Alice in Madison and flew back to California, where they had quickly booked a series of shows in Pasadena for the following week. My buddy Fred Coury, the drummer for Cinderella, was brought in to replace me. I remember going over the parts with him. I wanted to be gracious about the whole thing, and Fred was cool about the opportunity. He was well rehearsed and told me that *Appetite* was one of his favorite records to jam to.

We quickly sold out four homecoming shows at the Perkins Palace in Pasadena, and I was completely bummed that I couldn't be a part of it. I was angry too, at Slash in particular. I told him, "Dude, if you broke your finger there's no way you'd let them get someone else to sit in for you." Slash just shrugged, which is his standard noncommital response. Regardless, the shows went on, and Axl was cool about it, introducing me and bringing me out on-stage every night. I played tambourine on our new song "Patience." I'd talk to the crowd a bit and give Freddie major props. I'd say, "You rock my world, buddy." He was a good guy, but the situation just sucked.

# ⊱ NIKKI DON'T LOSE THAT HEARTBEAT ⊰

We were holed up at the Franklin Plaza suites again, and one night Nikki Sixx came over. It was me, Slash, and his new girlfriend, who I thought was a total bitch. Our drug dealer came over, and we got about a pound of coke. I grabbed my share and went back to my room to shoot up. After a few minutes, I decided to rejoin the group.

The door was open slightly, and when I pushed at it, it wouldn't open. "What the fuck?" I peeked in and discovered why the door was stuck. There was Nikki's huge, motionless body, passed out against the door with his face up. He had turned completely purple. "Shit!" With my shoulder, I put all my weight against the door and forced myself in. Everyone else was gone and Nikki was just lying there. I thought he was dead. I still had my cast on my right hand, so it was useless. With just one hand, I tried to drag him into the shower.

Suddenly, Slash's friend shows up, and I'm all, "Call a fucking ambulance. Call 911 now!" She just stood there. "Fucking call 911, you bitch!" She still just stood there. I swear she just wanted to be there when Nikki Sixx died. "Fucking help me drag him into the shower." Again, she had no reaction. I'm pulling, pushing, shoving, half dragging him with one hand, working as hard as I could. I made it into the bathroom and rolled him up and into the tub. I turned on the cold water and trained the flow from the showerhead directly on his face.

Nothing.

I started smashing him in the head, and I can still hear the sick sound it made when my cast slammed into his face. But Nikki wasn't moving. Not even a groan, despite the freezing cold water, the hammer blows, and me screaming at the top of my lungs.

*Zip. Nada.* Now, I was really starting to freak. I started slapping him in the face and suddenly, in what felt like milliseconds, the purple drained away from his face. It was the weirdest thing. Right away he came back to life. Ten minutes later, an ambulance came and the paramedics rushed Nikki down to Cedars.

The next day I found out he called someone to pick him up. He wasn't cleared to check out and had no business leaving. He called me and asked, "Stevie, what the fuck happened last night? My face is killing me."

I told him, "I was slapping you with my cast, dude, you turned fuck-all purple. I dragged you in the tub, got some cold water on you, and slapped you until you came to."

All he said was "Fuck. My head is killing me." He told me that as soon as he got home, he shot up another dose. I thought *I* was insane.

Years later, I found out that he's fathered several children, and saving him is one of the few things I feel proud about having done. Had I not been there, who knows? Everyone else had already left the room that night. And for all the things I feel horrible about in my life, this isn't one of them. I stepped up for my friend. Now, that friend's a dad; he's got a family, and I'm happy I was there for him. Plus Nikki is a legend, an amazing individual.

Stephen Pearcy, lead singer for the band Ratt, came over to the suites the following night. I had known Stephen for a good while. We used to hang out a lot at his house in Coldwater Canyon. I still had the cast on, but it was all wet and soggy. I was like, "Dude, help me get this thing off. It's driving me nuts." He helped me cut it open and remove it. God, it smelled so terrible, and Stephen said, "Jesus, Stevie, that fucking stinks like rotten meat." My hand was still badly swollen, and I had aggravated it to the point that I had to get a new cast put on. This prolonged the healing process, but fortunately, because of scheduling, it didn't force me to miss any other shows.

## ⊰ MOVIN' ON ⊱

The year 1988 rolled around, and we now had a gold record. Expectations were growing as we received word that sales were showing no signs of slowing down. I felt better once I got the cast off and buried my earlier feelings of insecurity by convincing myself there'd be no more fuckups.

The band got together to receive our *Appetite for Destruction* gold record plaques. I felt like I was floating on cloud nine. I wondered, "How much bigger can it get?" Immediately after the presentation, I visited Grandma Lilly and Grandpa Norman and gave the gold record to them. They were as much responsible for my success as I was; they were always there for me. I loved them dearly, and I wanted to show my appreciation. They hung the record above their TV, where they could always admire it.

I topped off my celebration by visiting my buddy Rob Benedetti at Sunset Tattoo. This time I wanted my very own Guns N' Roses tattoo. No skulls or guns or knives. I came up with this idea of a heart with wings on it. It cost me $145, and to me it meant the freedom and the love I've had with and through Guns N' Roses. That's how I felt, and it's what the band represented to me. Now, everyone in the band had their own custom GNR tattoo.

Rob mentioned to me that he also designed drumheads. In fact, it was Rob who came up with the idea of the little metal device that goes in the circle of the bass drum, where the mike is placed. It prevents the head from breaking or tearing. I was so happy with how the tattoo came out that I felt I was on a roll. So I had him make me those bass drum heads he was talking about, using the same design.

In mid-January I rejoined the band for a show at the Cathouse. There was no fanfare; it was just like rejoining your family at the dinner table. Ten days later, on January 31, we flew to New York, where we performed at a club called the Limelight. MTV had contacted our management about taping one of our live performances while in the Apple, and it was scheduled for our appearance at the Ritz on February 2.

Label mates Great White opened for us. After their set, it was time for us to hit the stage. I'm all ready to go, and fucking Axl is holding us up. Of all the times for him to do this. MTV was there, and this was huge, but eventually the MTV guys were like, "We gotta go, we gotta get this going, guys."

Axl's like, "Fuck it. I'm not going on unless I have my bandanna!" Apparently, he couldn't find it after tearing apart the little hovel they gave us backstage. Of course the rest of the band was avoiding any eye contact with Axl, preferring to wander off, out of earshot, to do their grumbling.

Finally, I couldn't take it anymore. "What's wrong with you, Axl?" He shrugged me off and continued with his insane tirade. He had all of our roadies looking around for people who had scarves or bandannas. I said, "C'mon, Axl, let's just go on."

He blurted out, "Fuck that. Fuck you. I need a bandanna or a scarf or I'm not doing this."

Now, we're thirty minutes late. The cameramen were tired of standing around and said, "We're outta here." I was the only one who was openly begging them to stay: "Please, don't go, we'll go on." I'm sure that's why I'm featured prominently throughout the video, because I showed some respect for the MTV crew.

Axl finally found a fucking scarf, some powder-blue, girly-looking thing, and the show began. He put it on, and he got this *Little Rascals* Alfalfa look going, because his hair was pushed up, like a ridiculous cowlick, on the back of his head. I'm sitting there playing and just laughing. "You dick, look at you. You couldn't go on without your scarf, and now you look like you're in an *Our Gang* movie." Someone must have tipped him off, because he finally got wise to it and adjusted the bandanna.

In spite of all the drama, the show went off fantastically. It's become one of the most widely bootlegged performances of the band. YouTube has the audio with stills, and a full video feed of it exists somewhere. It was also aired on MTV a bunch.

After the show I hung out with an old friend, Athena. Athena is Tommy Lee's sister. We had known each other for years. Back in '85, Slash and I were walking around Hollywood, by Sunset and Fairfax. Athena lived in this apartment building next to a health food store, and she and a friend were hanging out the window. They shouted out to us, invited us up, and the four of us got along famously. After that we would hang out, go to clubs, and invite them to our shows. They were the same age as us, and we had a lot in common.

Fact is, we've been friends ever since. I always thought Athena was such a beautiful woman. She was a skinny girl with great breasts. They were *so* big, she had to have a breast reduction because they were hurting her back. I would rate her as an easy *eleven*, that hot. She was such a sweetheart, a wonderful girl.

I was staying in a suite at a hotel that was in the same area where they did rehearsals for the famous *Honeymooners* TV show back in the fifties. After the Ritz concert, I invited Athena back to my room. We had a big pizza, and it was the *greatest* New York pie I'd ever tasted. We were eating, and I said, "I wonder if the Honeymooners ever did this?" I threw a slice of pizza against the wall. Then she threw one. It was a real mess. We were just laughing it up, having a hell of a time. We were drinking and one thing just led to another. Afterward, we just took in the sounds of the city around us.

We had made love in the spirit of fun. She had no one special in her life at the time, and although I was in love with Cheryl, it was very early in our relationship. Afterward, it was simply, "Hey, we did it, and that's it." If I was with one of the crazy babes who would fixate on us, it could have been awkward, but she was so cool. The urge just built up from knowing each other and being attracted to each other for a while.

There have been times when I thought I could marry Athena. But we were so good as friends, we both knew it was best to continue the no-commitment way. I don't think I ever considered the stabilizing effect a good woman might have had on my life at this time. Not a wife, but a levelheaded go-to girl who could have been in my corner with some good counsel from time to time. I think that Athena could have been that girl, but I was way too young and immature to even think about that at the time. Why have a level head when you can just have head?

## ⊰ HEATHER LOCKJAW ⊱

Around this time I had also been hanging with Athena's brother, Tommy Lee. He called me one day and said, "Dude, why don't you come over? We'll go dirt biking."

"Fuck yeah! Can I bring my little brother?"

Tommy said, "Sure!" I thought, "Jamie's gonna love this." He was just a kid, twelve or thirteen at the time, and a huge Mötley fan. We got in my Mercedes and cruised over to Tommy's house, which was in an upscale, gated community off Ventura Boulevard, way up in Woodland Hills.

Heather and Tommy were very much an item those days. But when we arrived, Tommy said she was sick. She had the flu or something. Eventually she came down and I introduced her to Jamie. It was clear that she was not feeling well, which is probably why she seemed a little bitchy, but I always smiled and treated her with respect. She was a gorgeous girl, but she was a TV star, kind of stuck-up. She was always worried about Tommy on the road, jealous about girls. Personally I knew she had every reason to be concerned.

Tommy had a couple of Hondas, and we took the bikes behind his house. There was a massive dirt mound just a short way back, perfect for riding. That day I showed Jamie how to work the brakes, the accelerator, and the clutch, and how to change the gears. It felt good being the big brother, showing him how to drive a motorcycle. He did great and it was a fantastic time. At the end of the day, we thanked Tommy and Heather for their hospitality and vowed to do it again sometime. Tommy just wanted to frolic and play all the time; he was the same fun-loving person I was. Two drummers who just want to bang away the day. Heather was definitely chaining him down, but guys like Tommy and me will put up with a lot and then, look out.

That weekend, a movie chronicling heavy metal's increasing popularity, particularly in L.A., opened. It was called *The Decline of Western Civilization Part II: The Metal Years*. As far as I was concerned, the best part about the film was that we weren't in it. Izzy played guitar and Axl did backing vocals, however, on the soundtrack's updated version of "Under My Wheels" with Alice Cooper. I thought that kind of rocked.

Most of the bands featured in *Decline* were struggling to make it but were overly confident about their shot at success. In the end, few of them ever truly made it. One scene in particular, featuring a shit-faced Chris Holmes, WASP's guitarist, was depressing as hell. He was sitting with his mother, cursing up a storm. He was talking about dying, complaining about being a rock star, and his mom was just sitting there smiling. I thought the whole segment was a real downer. There was, however, some comic relief in there for me. Particularly the scene with Paul Stanley lying in bed with a number of scantily clad young females. The scene looked so artificial. I couldn't help but think that after the shot somebody yelled, "Cut!" Paychecks were handed out, and everyone just went home.

On March 31, 1988, we did another acoustic performance, with me playing drums, on a show that was called *Fox Late Night,* a variety hour. It featured a black host, so "One in a Million" was out of the lineup. Instead we did a midtempo version of "You're Crazy" and also "Used to Love Her." When the host introduced the second song, he blew the surprise punch line of our tune by announcing: "Coming up, Guns N' Roses perform 'I Used to Love Her, but I Had to Kill Her'?!" The first time I ever heard the lyrics to that song was back at the Hell House. Axl and Duff came up with them. I thought it was so damn funny.

The reason I remember that performance so well is because we played "Crazy" the way it was always meant to be played: slower, sleazy, more bluesy, with much more feeling, and not the frantic sped-up version on *Appetite.* Even though Axl had to censor himself for TV and leave out all the "fuckin's," he did a masterful job, and it's definitely my favorite rendition of "Crazy." Check it out on YouTube: http://www.youtube.com/watch?v=RuqUqNuh2nQ#.

### ⫷ PARTY ON ⫸

Of course, life was a nonstop party. Axl would have Del and Wes out with him. Duff and Slash were a little less distant toward me since my hand had healed, but honestly, who needed the fucking drama? I didn't, so I decided to hang with Tom Mayhue, my newly appointed tech. He had already been on many tours, particularly with Dokken, serving as Mick Brown's tech. I was getting massive amounts of coke, but I found that I would burn out on it from time to time. So, on many occasions, I'd give what I had left to Doug and just be like, "Do whatever you want with it, but just get it away from me."

The time had arrived for us to shoot our second video, "Sweet Child O' Mine." It was filmed at the Huntington Ballroom in Huntington Beach. All the guys got their girlfriends to be in it with us. Axl had been dating Erin Everly for a few years. She was the daughter of Don Everly, of the Everly Brothers. She was just the sweetest girl, and I thought for sure that she wouldn't be able to stand Axl for long. They were always arguing about this or that, and at times it would get really intense.

When he first brought her around, she was so cool and I honestly thought, "I hope Axl doesn't screw this up." Duff was dating and became engaged to a girl named Mandy who performed in an all-girl L.A. band called the Lame Flames. Slash had his current girl there, and I brought Cheryl in. A friend of mine worked with the L.A. punk veterans TSOL, with whom we had played many shows, and asked me if I would wear the shirt in the video. Why not? I happily obliged.

A few weeks after the "Sweet Child" video was shot, our record was certified platinum. It was an unbelievable achievement for the band and our whole team. Our popularity was soaring to the point that we were being referred to as a "supergroup" in the rock rags. We truly were the rock stars we'd dreamed of becoming, and it was finally starting to sink in. But it was soon plain to see that the bigger we got, the colder we became.

## ⫷ CLINT AND MY FIRST OD ⫸

Right after the "Sweet Child" video became a sensation on MTV, the band's growing popularity came to the attention of screen legend Clint Eastwood. Someone must have suggested bringing us into his new film project, *The Dead Pool.* Around the same time, I was asking Dougie, "Dude, can you get us in a movie? Or a TV show? A fucking cartoon? Anything?" So it was a pretty cool coincidence how it all worked out.

Axl would later proclaim the movie to be "a piece of shit," encouraging fans *not* to see it. It featured a would-be Axl-type rock star, Johnny Squares, played by a then-unknown Jim Carrey, lip-synching to "Welcome to the Jungle." In the film's plot, he is murdered in what would be made to look like an overdose so that participants in a pool for betting on when celebrities would die could benefit from the bets they had placed in their morbid game.

We were scheduled for two days of filming. The first was at Forest Lawn Cemetery in San Francisco, where we were cast as rock 'n' roll friends of the deceased Johnny Squares. We all knew this would be great visual exposure. We were going to be in a movie. It was exciting and even Axl showed up on time.

During the first day of filming I hooked up with this stripper chick who was an extra on the set. She was a local, so we went back to her place. We smoked a little weed and she made us tea. I noticed that she had slipped a brown powder substance into my cup. I assumed it was some kind of spice or something. The last thing I remember, she had my head between her legs and was forcing my face against her groin. "Now I'll show *you* how to eat pussy," she purred. She was in control and I was floating, floating away on some silken cloud without a care in the world.

I began to chuckle at her remark as those clouds enveloped me, and then *nothing.* I passed out while she was mounting me. What a time for my first heroin overdose. I woke up in a hospital room the day we were supposed to be filming our second scene for the film. I had no idea how long I had been out. In fact, I had no idea where I was or what had happened, but as my vision cleared it was apparent someone was keeping vigil over me. Someone was at my bedside patiently waiting for me to come out of it, though no one knew if or when that would be.

I blinked. I blinked again. It was Axl. Axl got up and was now standing over me. He smiled. He looked genuinely relieved. He said,

"Man, that was close, Stevie." He was the only one there. Later, a nurse told me he had sat by my bed the whole time. The other guys went ahead to do the movie but Axl stayed at the hospital.

What Axl did was so noble, so selfless and surprising, that I cried, and they were tears of joy. He kind of shifted uneasily when he saw the waterworks, but that was just too bad. I felt nothing but unbridled love for him at that moment and I didn't care if he saw it.

There he was, blowing off a chance to be in a scene for a major movie release, to stand by me, his own personal vigil, just Axl. Goddamn Axl. Soon as I thought I had him pegged, he went and did one of the most touching, meaningful things anyone's ever done for me. That was so cool. Thanks, Axl.

Here's the thing about Axl. He demands emotion. "Love me, hate me, but don't you dare fucking ignore me." He will not tolerate a vacuum. Sometimes I think that's why he would keep our fans waiting for three hours before going on. He demanded an emotionally charged atmosphere at all times. He wanted a life spent on the frantic jagged edge, and that's why he could deliver that unique urgency in his lyrics: he lived it.

Having Axl there really had a powerful effect on me. I was back on my feet in no time. The nurse called it "the power of youth." More like the power of stupidity. Anyone with just an ounce of sense would have known to wait for a clean bill of health from the doctor, but not me. As soon as I could get out of bed without falling over face-first, and that happened at least a dozen times, I was leaving. Finally, I ripped the tubes out of my arms and just stumbled out.

## ⇥ THE MAIDEN ⇤

Next up was a tour with Iron Maiden. I dug the band; I remembered jamming to their *Piece of Mind* record over and over when I was staying at Brad Server's back in 1983. I was hanging with Nicko McBrain, Maiden's drummer, one night in Quebec, a beautiful city. We were checking the sights, having a few drinks, and Nicko starts talking shit to me. I didn't say anything to start anything, and I didn't say anything *back* to him either. He was so drunk and he was getting pissed at me for no reason at all. I don't know why. He is one big wide-body dude, and I thought he was gonna kick my ass. Luckily he drank himself into a stupor and got all pie-eyed and quiet. I just slipped away and grabbed a cab back to the hotel.

I do remember one time during this tour when probably a couple of hours after the show had ended, I walked into the greenroom, where about twenty fans were waiting around to meet the bands. Like I said earlier, the guys in my band were not into meet-'n'-greets, record signings, nothing like that. So I walk in, and it's completely silent. All the kids looked bummed out; they all had their heads down, a real sad scene. So I go, "Hey, c'mon, this is a fucking party!" Suddenly everyone looks up at me, and the place erupts. It felt great. A big "*YAAAYYY!*" filled the room. One of the guys had a few joints and I lit one up after another and passed them around. I talked to everyone and signed autographs. I was so surprised at what a difference a little affection could make. I mean, as far as lifting these kids' moods it was like night and day. It was really rewarding for me, and I always wished we had done more shit like that as a band.

The Maiden tour brought us back through Canada, back to the U.S., and ended in Sacramento, California. Our gear was set up, and about two hours before we were to go on, I hear that Axl can't do it, his vocal cords are shot. Although a lot of our fans were going to be upset, the band was aware of Axl's true intentions and understood what was actually going down, as you'll see soon.

So at the last minute, local Sacramento band Tesla filled in. I think they even used our equipment. I was disappointed because the next show was going to be back at the L.A. Forum, a place where I saw so many bands when I was growing up. "Aw, man. I always wanted to play the fuckin' Forum." Our buddies in L.A. Guns got the gig, and they continued opening shows for Maiden.

At Irvine Meadows Amphitheater, all of us minus Axl got up on-stage with L.A. Guns and performed two songs to an ecstatic audience. We had bowed out of those shows at the last minute last time, and many of the kids went specifically to see us, so we felt obligated.

## ⊰ DREAM COME TRUE ⊱

Just a few days later, our dreams came true again when our long-awaited tour with Aerosmith kicked off in Illinois. I remember there was this one-way road to the venue that went on for miles. So we took a helicopter from the hotel right to the backstage area. It was so cool, so rock 'n' roll. After the show we were in the dressing room and all the guys in Aerosmith came in, Tyler, Perry, Whitford, Hamilton, and Kramer, and said, "You guys rock. You guys definitely rock." That was the first time we met them, and it couldn't have been any better in a dream. Our heroes telling *us* they liked the way we played. And one of the main reasons they were impressed was because Axl's voice was fucking incredible, godlike in its range and intensity.

That was because he had dropped out of the end of the Iron Maiden tour to give his voice a good rest. You see, Aerosmith meant so much to him, and so much to us, that he didn't want to blow out his voice. He wanted to be well rested. And hearing Joe Perry tell us we kicked ass made it all worthwhile.

There was one show we played, I think it was in Indiana, where they even sold out the seats *behind* the stage. We definitely brought *a lot* of those people in. Everybody was singing along with the songs with a ton of energy. There was so much excitement, I had to yell to Tom as loudly as I could, "Get me a bucket. Get me a bucket!"

I must have sounded like I was in a damn Monty Python movie. Playing the show felt so great it actually made my stomach turn cartwheels and suddenly I had to vomit. As soon as the song ended I puked right in the thing. After I shed the jitters, it was even more fun than I could ever have imagined. I was like a little kid up there, sneaking under my own tree on Christmas day.

Aerosmith are my heroes. Period. I respect Steven Tyler more than any other front man in the world. He really was the coolest, greatest, most down-to-earth guy. Besides the fact that he was an amazing performer, and a rock legend, he was truly the nicest, most genuine guy. Remember, he's a Tallarico and I'm a Coletti, so we share a little linguine lineage too.

The management told us to keep the drinking and drugs out of sight, as the boys in Aerosmith were all recovering addicts. After some of the sound checks in the outdoor theaters, Tyler and I would go out in golf carts and race around the venue.

He told me some amazing stories about his battles with drugs, how in the old days he would be dancing off to the side of the stage and he had his assistant there, ready with a syringe all filled up. He also confessed that in his famous black-and-white-striped outfit, he had a mini-pocket inside the scarf where he stashed quaaludes, Valium, Percodan, a fucking pharmacy in his folds. Then he looked me right in the eye and said, "Don't let *yourself* get too mixed up with that shit."

I looked him right in the eye. "No worries, Steven, I won't." Ha. *Famous last words* . . .

One night Steven and I brought about ten girls back with us to the tour bus. We told them to get naked. Steven assumed the role of a director. "Now you three, suck his dick. You, sit on his face while he eats your pussy. You two, make out." Steven Tyler, rookie porn director of the year. And me, the new king of porn. That was the most memorable night of the tour and one of the top ten memorable nights in my *life*. Tyler is nonstop fun.

We got to New Jersey to play at Giants Stadium, and after an amazing show, I went back to the hotel really looking forward to seeing Cheryl again. I began having some serious thoughts about our relationship because I figured I had gotten myself pretty well set up with a kick-ass successful rock band and had good money coming in as a result of it. And I don't know, maybe the heart to heart with Tyler from the other night was making some headway into my thick skull. Maybe it was time for me to settle down a bit.

So as soon as we met in the lobby, I took her to my hotel room and said, "Take off your clothes." We made love like never before. She could sense my passion and responded with just as much enthusiasm. We lay in bed afterward and as far as I was concerned, this was *the one*.

# 15

# TRAGEDY *and* CONTROVERSY

## ⊰ DONINGTON DEBACLE ⊱

In the middle of the tour, we were flown out to England to perform at the annual Monsters of Rock festival at a racetrack in Castle Donington, England. It was August 20, 1988. Opening the show was a German metal band, Halloween, then us, then Megadeth, David Lee Roth, Kiss, and Iron Maiden. To get us there quickly, a Concorde, the world's fastest commercial jet, was chartered. It took only three and a half hours to cross the Atlantic. A normal flight would have taken like eight. The entire cabin was first-class, prime rib, Sennheiser earphones for the sound system, and your own gift bag when you got to your seat. I looked out when we were at our highest cruising altitude, like sixty-five thousand feet, and I could see blue, dark blue, then indigo, then if I craned my head to look straight up, black! Also, if you looked straight out, you could see the curvature of the earth, just slightly, but it was there. Pretty fucking cool.

We arrived the day before the show. Lars from Metallica was there and we were snorting *anything* that looked like powder. We were doing this pink shit, I don't know what it was. It could have been crushed-up baby aspirin for all we knew.

Lemmy was there too, and he had a pile of speed on the table about a foot in diameter. All he did was speed. He offered me some, and I just did a little, a real tiny bit. It felt like fiberglass going up my nose. Dave Jr. (that's what we called Megadeth's bassist Dave Ellefson) and Dave Mustaine were there too. We had partied many times prior, including smoking up a little heroin.

In the middle of the afternoon we hit the stage. It was a madhouse. Over a hundred thousand kids were cramming against the front. The racetrack was selling these big thirty-two-ounce beers. The kids were drinking, and they weren't about to go through this whole fucking crowd just to urinate at a stall, so they pissed in the bottles. Before we went on, we were standing at the side of the stage looking at the size of the crowd.

Suddenly, we saw what looked like a swarm of giant locusts flying through the air; they were actually hundreds of these plastic bottles of urine soaring over the crowd. We were like, "What the fuck?" *Bam, pop!* People were getting hit in the head and splattered with pee. But it wasn't going to change anything. We had gotten spit on, we had bottles of booze and beer thrown at us, and we had gotten in shoving matches with fans and other bands, so what's a little projectile piss?

I was surprised to see so many Guns N' Roses banners waving in the crowd. By the time we went on there were 120,000 people screaming and jumping up and down. It was really an impressive sight for us all. Everyone was so out of control, and we had to stop the show several times because people kept rushing the stage. Axl asked the crowd to settle down and back up. People were getting crushed at the front of the stage. It wasn't until the next day, after we flew the Concorde back to the U.S., that we were told that two kids were killed during our set. They were trampled to death.

I was shell-shocked. Numb. I couldn't believe it. Of course, the media blamed the band, fueling our notorious bad-boy image. And we were just starting to get a broader, more friendly public image going when this happened. This was partly because "Sweet Child

O' Mine" had broader appeal than "Jungle" as a hit. We tapped into a larger following with that tune, reaching more mainstream rock- and pop-minded folks.

I called my mom and told her about the terrible tragedy. I never stopped to think of why I called Deanna instead of Cheryl or Big Lilly, but I immediately felt some solace as soon as I shared this horrible news with her. She was shocked but didn't unravel. She managed to be very compassionate and real with me, explaining that I wasn't to blame. She reasoned that the promoters have to control the numbers and the way the seating is set up. I understood, but it didn't make me feel a whole lot better. I felt like I had somehow been a cog in some bigger machine that hurt those kids. It was weeks before I felt anywhere near normal again. I let Mom know that I would be back in town soon and would call her. To this day, the Donington tragedy still haunts me like a waking nightmare.

## ⊰ FAMILY MATTERS ⊱

We had two shows with Aerosmith at the Pacific Amphitheater in Costa Mesa, California, during September. I invited my family to the first show, but then I guess they got the idea to just come by and surprise me at the hotel beforehand. After all, they hadn't seen me in a year. Me and Ronnie had been up all night doing coke, a lot of it. I had a huge pile of krel on this chest of drawers that was set up like a table. We were tweaking hard. At eight in the morning there was a knock at the door.

Consumed with paranoia, I asked Ronnie, "You expecting anyone?" He shook his head no. Another knock. I slowly stood up and made my way to the door. "Hey, Stevie, open up." That voice, I know that voice. It's cool.

So I opened it, and standing there were my mom, dad, and little brother, all sporting huge smiles. I was horribly, unreasonably pissed. "What the fuck are you doing here? Don't just come over unannounced." In an instant, I just saw their happy expressions turn to disbelief, then horror. Fuck 'em. I slammed the door and tried to get a little shut-eye before the show. Let's face it, I was such a dick, I still feel bad about it. Hey, Ma, you out there? I apologize for that.

The next day we performed our last show with Aerosmith. As the crew was setting up the equipment I ran into an old bass player friend of mine, who I had set up with tickets and passes. He mentioned that he was going to score some dope, and I said, "You know, I got twenty dollars. What the hell, pick me up some." I unzipped the fanny pack around my waist—I called it my "hippie" pack—and handed him a crisp twenty. I had done the shit only a few times up to this point; it wasn't like I had a connection or anything. After the show, about three hours later, he came back and gave me the dope. I ripped a piece of foil off from a catering table and went to our bus. I went straight to the back for some privacy and smoked that bad brownstone.

## ⚔ KARMA KILL ⚔

I was feeling all happy-go-lucky, high as a kite. I was on my way to party with Aerosmith. I was smiling, on top of the world as I swaggered toward the greenroom and opened the door, striking my best rock star pose and surveying the scene. There, just five feet ahead of me, was Steven Tyler. He was smiling, chatting with someone. He turned to me, looked in my eyes, and his smile faded. He just *knew*. He shook his head like "Oh, man" and looked the other way. I didn't understand it at that precise moment, but he was sad for me. Then it sank in and my own smile faded as quickly as my family's had in the hotel.

I was dumbfounded. I turned and quickly walked away. Not being cool with Steven Tyler is as uncool as you can get. He got to know me, but I was only playing along, only saying the things he wanted to hear so that he'd think I wasn't doing that shit. He told me about how he was a loser for doing it. And there I was with those damning pin-dot eyes, busted. I felt as though I had let him down far more than I had ever let myself down. Steven Adler, the fucking fool, had deceived a man he loved and admired. I was miserable for days; I wanted to blow my fucking brains out to stop the pain.

In my head, I've identified that moment as the definitive turning point, the precise moment when things began to go from occasionally off-kilter to very darkly wrong in my life. Breaking my hand was the first warning shot, but I just kept going, gathering momentum, blowing right through those sawhorse barriers on life's highway, keeping my metaphorical pedal to the metal until I ran out of road, crashed and burned.

## ⇤ FALL FROM GRACE ⇥

The day after our last show on the Aerosmith tour, Dougie called and said he had been asked if I'd be interested in a video shoot. Comedian Sam Kinison was super-hot at the time, and he was making a music video to promote his new album. I had known Sam for some time. A few months earlier, I had taken a number of friends to see him at the Comedy Store on Sunset. He invited us backstage and cut thick lines of coke for us all. He was a wild one, one of us, no doubt about it.

The video was for his outrageous video parody of the song "Wild Thing," a hilarious retooling of the Troggs' classic from the sixties. His idea was to have a big party, invite all of his rocker friends over, and double-duty the gathering as the cast and setting for his video. Slash and I got Sam's invitation and went together. When we arrived, we were excited to find ourselves in amazing company. Our dearest buddy from the Crüe, Tommy Lee, was there and Bon Jovi showed with his entourage. I just looked at all these fun-loving rockers around me and said, "Fuck yeah." It was like some perverse validation; I was so proud to be included in this group of genuine rock stars.

Fresh from the massive religious scandal involving televangelist Jim Bakker was the video's sole female star, Jessica Hahn. She had single-handedly brought down that hypocritical Bible-thumper Bakker and his mascara-streaked wife, Tammy, and was now Sam's current slutty girlfriend. Wonder how that played out on the religious right? Fucking hypocrites.

I thought Jessica was a pig. She had on so much makeup, she looked like a mannequin. But Sam adored her, so we went along and treated her sweetly because he was our hallowed host and friend. Her huge, rocket knockers really turned me off; they kept threatening to poke out our eyes. Sam was rolling around with her in this little love pit and everyone was pointing at them, cheering.

They had a keg of beer for us and we were all quaffing with a vengeance. As they shot the scenes we weren't in, Slash and I camped out at the tap and became mightily shit-faced. I don't even remember the shots we were in but I do recall that later, they insisted on doing some additional pickup shots, including one of a drunken Slash clumsily falling into a trash can. The video was in regular rotation on MTV through late 1988 right into 1989. It was a great idea and a great time.

The annual MTV Awards were held that year on September 7 at the Universal Amphitheater in Los Angeles. There was no question that we were the hottest band around. We were up for the Best New Artist award, which is now universally regarded as the "kiss of death award." So many bands have gotten that award only to crash and burn on their sophomore album. They should outlaw that award. It's like being on the cover of *Sports Illustrated:* instant jinx.

At this time, Slash had been hanging out with the notorious porn star Traci Lords. During the ceremony, when the envelope was opened and we were announced as the winners, the producer decided to have Slash and Traci be the ones to accept. As for the rest of the band, we honestly didn't care who went up there as long as it was one of us.

"I'm Traci Lords," she announced.

"I'm Slash," he mumbled.

Traci continued. "Guns N' Roses is very happy to accept this, thank you [giggle]." How profound. Later that evening we performed "Welcome to the Jungle" live for a less-than-enthusiastic audience. These days, MTV has smartened up; they pack the front of the stage with wild and crazy fans driven to delirium for the cameras. But back then, they set it up much like the Academy Awards, with all the biggest stars up front. At the end of the song, I threw one drumstick out as hard and as far as I could; the other I gently tossed right to Steven Tyler, who was sitting in the front row, hoping it would raise a smile. Tyler didn't flinch, and the drumstick just lay on the floor with no one giving a shit. That hurt, but it didn't surprise me.

After we performed, the host, Arsenio Hall, cracked a joke about drummers: "I'll never understand, they throw their sticks into the audience, so there's a big fan out there with a stick in his eye going, *'I love them guys. I love them!'*"

On September 17, 1989, we played the final show of the *Appetite for Destruction* tour. It was a big festival outing in Texas that also featured Australian pop icons INXS and reggae artist Ziggy Marley, the son of the legendary Bob Marley. We flew in the day before the concert. I recall sitting in my hotel room watching cartoons when I glanced out the window. There were a lot of big Jamaican Rasta guys walking around, and they looked kind of threatening, but that was just my sick head.

My eyes drifted to the pool, where Michael Hutchence, singer for INXS, was regally sitting. He was with a beautiful girl who looked like a supermodel. They were lounging in pool chairs, chatting with each other. I thought, "How cool is this? I've seen this guy in music videos for years, and here I am about to play a big concert with him."

The show was the absolute *worst* we ever played. For some reason, the guys just weren't into it and the reason was simple: they wanted to go home. I realized, "Hey, *I* want to go home. We've been on the road for ten lifetimes and it's time to shut it down for a while."

To add to our misery, it was raining that day. We were in Texas Stadium, a partially covered arena that had a huge opening over the playing field. From the stage, I could see rain pouring down on the crowd, but we were kept mostly dry, except when it would get gusty. It was the weirdest-looking setup. We played our set in record time, just wanting to get it over with. "Sweet Child O' Mine," our mid-tempo hit, was practically played at twice the speed.

Christ, we had actually been on the road for two years. It was time for a break. Sure, it was a really big show, but I *never* once thought that there wouldn't be hundreds more big shows in my future, so there was zero nostalgic wistfulness as we bowed and got the fuck out of there.

Once back home, I enjoyed my much anticipated time off with Cheryl. She was with me every minute. She came over and from the moment we hugged it was quality time and quiet time. Fresh sheets, pull the blinds, we're sleeping in. Looking back, at least up until that point, it may have been the happiest time in my life. I woke to Cheryl's heavenly smell each morning and fell asleep to it each night. I finally had time for the love of my life and the life that I love.

Imagine having a secure, steady flow of income in your early twenties, becoming famous doing what you love for a living, and spending each day with the woman of your dreams while not having a care in the world. You can't beat that. But you can destroy it.

I was experiencing things in life that were beyond my wildest dreams, like when we made the cover of *Rolling Stone*. But we were just taking it all in stride, like it was no big deal. I was wearing a shirt that Tom Mayhue had given me by way of Mick Brown from the band Dokken. Dokken was on tour with Van Halen, Metallica, and the Scorpions. It was called the Monsters of Rock tour (not to be confused with the Monsters of Rock festival in the UK). The shirt was a brilliant parody of that, a sarcastic play on words dubbing it the "Hamsters of Rock."

A few days after the *RS* cover hit the stands, I was in the shower when Cheryl came running into the bathroom. "Steven, guess who just called?"

I shut the water off. "Who?"

"Eddie Van Halen," she said.

"No shit! What did he want?"

Cheryl said, "He was pissed. He was like, what is this shit with you wearing '*Hamsters* of Rock'?"

I couldn't believe Eddie would take something so minor so seriously but was nevertheless very excited that a rock god had called me. "Oh, no way . . . Did he leave a number to call him back?"

"Nope," she said, "he just hung up!" I had heard that Eddie's temper flared when he drank, so I let it roll off my back pretty quickly.

In late November, our EP *GNR Lies: The Sex, the Drugs, the Violence, the Shocking Truth* was released. The cover was a send-up of English gossip rags, labeled with all sorts of sensationalistic, over-the-top headlines, like MAN SUES EX-WIFE, "SHE TOOK MY SPERM WITHOUT PERMISSION" and SEVERED HEAD FOUND IN TOPLESS BAR.

Again, I was given a few photographs to choose from that would be featured on the cover. The package included three new songs, an acoustic remake of "You're Crazy," and a rerelease of our original *Live?! Like a Suicide* EP, which on the vinyl and cassette releases was the G side (aka the A side). The B side was labeled R. We recorded the new tracks at the Record Plant recording studio off Sunset by Paramount Studios. The entire process was done over a single weekend. I played on three songs and didn't stay a second longer than I needed to. I couldn't wait to get back to Cheryl and party with her.

Now, I regret how quickly I took off. I think I missed out on some special band moments. But there's no way I could have known that at the time. To me, my work was complete, done. I never left the studio without everyone signing off on my work. It's just that some songs, like the single "Patience," really needed no percussion whatsoever. I was familiar with the song, and I actually thought it was a great idea to keep it strictly acoustic with no drums.

That's how I am with a lot of stuff. Pretty mellow when it comes to artistic interpretation. No ego with the drums. If they don't belong, they don't. Can you imagine the Stones' Charlie Watts trying to wedge a drum part into "Lady Jane"? I was fine with not playing the drums on every track until GNR said I couldn't play the drums on *any* track . . . ever. But more on that in a bit.

"One in a Million" featured the wildly controversial lyrics about "police and niggers" and "immigrants and faggots." I thought that it was a great song that needed strong words. It expressed a heavy sentiment that had to be delivered with no punches pulled. I knew that the words weren't directed to the majority of blacks, gays, or immigrants. It simply described the scumbags of the world. That's what Lennon did when he wrote "Woman Is the Nigger of the World." "Nigger" to him meant "slave." And we meant "lowlife." The song explained the shit that Axl, a naive hick from Indiana, had gone through.

Nobody thought twice about it, not even Slash, although I later learned that his mom was offended. I thought when Axl sang, "Immigrants . . . come to our country and think they'll do as they please, like start a mini-Iran," he meant, "Look how fucked up Iran is. Don't bring that shit over here." That was my interpretation. Also, because we loved that song, we wanted it to get a lot of attention, and this was the way to fan the flames.

The only shows played in support of *Lies* were in Japan and Australia. We had only had a ten-week break since our last concert opening for INXS at Texas Stadium outside Dallas. Japan was awesome. Especially for a blond like myself. I got the impression that they just worshipped blonds over there by paying them special attention. When we arrived at the hotel, I went straight to my room. I opened my bag and threw everything all over the bed like I always would. We went out on the town for a little while, and when I returned to my room, I was pleasantly surprised. I discovered that all my clothes had been folded so neatly, so perfectly. I didn't even want to touch them and ruin the artistry. I went, *"What the hell?"*

Then I heard the cutest-sounding giggles. There was a petite Japanese hottie hiding behind the curtain. She had her hands over her mouth, suppressing her excitement. Then I heard laughter from within the bathroom. There, another young girl was hiding in the shower. Together, they welcomed me to their country in their own special way. Japan: Land of the Rising Buns.

The *man* in Japan, the top promoter and a formidable leader in numerous business ventures, was a distinguished gentleman named Mr. Udo. He would take our entire crew out for fabulous dinners.

The lighting guys, the sound techs, everybody benefited from his extreme hospitality. On one such occasion, they brought us twenty entrées. They set a bowl of soup in front of me. I said, *"Domo arigato,"* which means "Thank you very much." They all smiled and bowed respectfully. Even the band was impressed. This, however, was the extent of my grasp of the language. I learned it from the Styx song "Mr. Roboto" in which they sing, *"Domo arigato,* Mr. Roboto." I looked at my dish, and I was taken by the sight of many tiny fish swimming around in the broth. It threw me for a loop, but I didn't hesitate to indulge.

## ⊰ FOOD FROM HEAVEN ⊱

The absolute best meal I had there was Kobe beef. It was explained to me that Kobe beef was made from cows that were fed copious amounts of beer their entire lives. They were raised to have soft muscles by being massaged for hours a day. I was told that when they kill it, the cow is pretty out of it. This meat is so tender. It's cut into little squares. They put a steaming hot rock in front of you, and you set a piece of meat on it, cook it, flip it over, and do the other side. It just melts in your mouth, definitely the best meat I've ever eaten.

We traveled on the famous bullet train, where they had the best food in their dining car. During our journey, I met the granddaughter of the Kawasaki motorcycle mogul. She was beautiful and nubile. I felt that when I was on the road, it was okay to play. So I charmed her and managed to get her to make out with me.

## ⊰ FAN-ATICS ⊱

The fans in Japan were crazy. I have a picture of Dougie pulling me through a crowd of hundreds of young Asian girls. It looks like something right out of the Beatles' movie *A Hard Day's Night.* There's one scene in that film where the guys are running down a street, chased by hundreds of fans. George Harrison trips, causing him and Ringo to fall. It was the funniest thing, but you could tell they really fell, no stunt bullshit.

Well, Duff and I experienced something similar. We walked out of the hotel one night, heading over to the Hard Rock for some dinner and drinks. We avoided the main exit and slipped out a side door. Walking around the block, from fifty yards away we could see about eighty kids waiting for the band, armed with records to be autographed. We had been signing shit all day, and I asked Duff, "You really wanna do this?" We just looked at each other and started running as fast as we could on the other side of the street, hoping not to be noticed. We were spotted, however, and every one of those kids started chasing us. We rushed into the Hard Rock and the security people there stopped the kids in their tracks. We sat down to food and drinks while the kids continued screaming for us outside. We waved at them through the restaurant window, not taunting, just friendly, and they went happily nuts. It was just another rock star experience.

We asked where the best nightclubs were in town. I wanted to go dancing and drinking. Many told us that the Lexington Queen was the place to be, so we took the advice and ended up there one evening. The owner of the establishment was a gay Asian guy who paid me special attention. It was a little embarrassing because he just *loved* me and treated us so well there. I was on the dance floor grooving with a few Asian girls, and I set my leather jacket down. After a minute or so, I went to collect my jacket, and it was gone. I ran over to the owner and said, "Hey, I was dancing here and someone swiped my jacket."

He told me, "It must have been an American model. Asian people would not steal from you, Steven." He said, "Wait here." He came back with a brand-new jacket with the logo for the Lexington Queen embroidered on the back. Later, I gave that jacket to my dad, Mel, and he really loved it.

here were many American models hanging out in Japan. I learned that they would work there for a few weeks at a time. Their employers would put them up in their own little apartments. So naturally, I hooked up with one. A tall, thin blonde; nothing too special. She took me back to her place. We made love and I really thought nothing much of it. The very next day, Axl meets the same girl. She turns out to be a total troublemaking bitch. He fucks her and she starts telling him that I was talking all kinds of shit about him. Why would I share negative stuff about him with some random girl I didn't even know? Axl was my brother and we were conquering the world together. If I had an issue with Axl I got in his face. That's the way I rolled. Always.

So Axl comes up to me and says something like, "This here is my woman, and she told me that you said I'm an asshole."

I said, "Your woman? You just met her, Axl. We fucked last night. That's all. I didn't say shit to that bitch." The argument just kind of fizzled out at that point with Axl mumbling something as he walked off. He was usually okay as long as he got the last word, whether you could hear it or not. Unfortunately, incidents like this only served to weaken my relationship with Axl.

# 16

# SHOOTING VIDEOS *and* HEROIN

The shows in Japan were amazing, all sellouts. The Japanese created their own colorful posters announcing the concerts and the fans were great. Like the Germans, they knew all the lyrics to every song. On December 4 we played NHK Hall in Tokyo; the next day we played Festival Hall in Osaka, then went back to Tokyo for three shows starting on the seventh and wrapping up on the tenth. For that last show, we made a dream come true for the band. We played Budokan, a world-famous venue where everyone from Dylan to Clapton to Cheap Trick had played. Cheap Trick's live album *At Budokan* was recorded there and we all loved that album. It was one of Slash's and my first records. We must have listened to it a thousand times when we were kids. Axl even mentioned that it was an influence on us all during the performance. I just remember playing and looking out over the crowd thinking, "Wow, this is where all those great songs went down."

We were exhausted, and Axl's voice was raw, but we rallied because it was our final show. Axl had actually apologized for "playing like shit" the night before at NHK. I got to do an extended drum solo during "Rocket Queen," and we closed with a fucking epic version of "Paradise City."

Tours in Japan usually lead to Australia, and that's what ours did. Three days after Budokan we performed the first of two shows at the Entertainment Center in Melbourne. It was a huge outdoor arena. The first performance was a sellout. The second was at about two-thirds capacity. I recall those shows fondly because I was able to hone my drum solo until it sounded really tight, light, and playful at first, and then very explosive. We never really planned stuff like that, and I think the solo just grew out of the middle of the song where Duff slapped a cool bass riff and I followed with a flurry of drumming. No one broke back in, so I kept playing, and each performance I'd carve out a little more solo time. It was a natural, totally spontaneous development, and I smiled ear to ear.

We also played in Sydney on the seventeenth and then were off to Auckland, New Zealand, for our last show of 1988. We performed that show in support of *Lies* and ten thousand screaming kiwis loved every minute of it.

I spotted the then eighteen-year-old supermodel Rachel Hunter hanging out with some friends on the side of the stage. Everyone was intimidated by her beauty, literally afraid to approach her, but I didn't care. I had the smile and the charm and was armed with the fact that I was in the best rock 'n' roll band in the world. I just went right up to her after the show and struck up a conversation. I knew I had to spend some time with this knockout, invest in some one-on-one chatting. We went back to the hotel together and talked all night. The next morning we had breakfast and then she showed me around some cool spots in Auckland. It was such a sweet, effortless time, and she was so gracious and hospitable. We never got past holding hands, hugging, and a little kissing, but that was just fine. I was in heaven and she was one helluva great kisser too.

When we flew back to L.A., I noticed more than ever how popular we had become. At the airport we were mobbed for autographs. People recognized us everywhere. A lot of folks would just stare and whisper. I guess they felt weird or too shy to actually approach us. I got used to it quickly. Everywhere I went, someone knew my name. At my favorite hangout, the Rainbow, the guys treated me the way they always had, and that was great. This was a place where I had spent a lot of time. I had partied at each booth and in every chair, and I had slipped away to fuck everywhere in the place. Different chicks had ducked under every table to give me a blow job. After the band made it, I felt especially proud just hanging out there.

So many people who hung out at the Rainbow looked and acted the part. They all had the rock star thing going. But could they really play? Could they really make it happen? We proved ourselves. I would always hug and chat with the owners Michael, Mario, and Steady. I asked Michael, "Hey, if I brought something in, would you hang it on the wall?"

"Sure, Stevie. Of course." I gave him a signed snare drum and a framed picture and felt honored to have something up on the walls with all the other pieces of rock star memorabilia. The Rainbow is like L.A.'s version of the Rock and Roll Hall of Fame, and to this day, my picture and drum hang on the ceiling.

After our last shows, we returned home for an indefinite period. We had no solid plans for the immediate future, no itinerary. Tama drums had signed me to an endorsement deal and they flew Cheryl and me out to Philadelphia, where I would make an appearance and do a photo shoot for drum advertisements. We spent a few days there with Cheryl's family. They were just the most down-to-earth, caring, blue-collar folks, and I loved them.

When we got back home, Cheryl and I retreated to my apartment, which was located by my grandparents' place. I didn't visit them much, it was just where I happened to be renting at the time. I was just partying, doing my own thing. I befriended this kid who lived across the street, a hippied-out stoner dude with shaggy hair, about twenty years old. I'd give him twenty or forty bucks and he'd pick up some heroin to smoke. At first, one run in the morning and one at night was fine. It was perfect. I had my waterbed in the living room, and I would just lie there all day watching TV. But after a couple of weeks I was shelling out $150 to $200 so he could make three or four runs a day. Hmmmm.

Cheryl was pretty oblivious to it all at first. She didn't party with me. She'd often leave around noon to go out shopping or hang with her friends. She'd come back before dark and I'd be lying there stoned out, high and numb to the world. She'd give me this big smile and show off what she got at the shops. I'd pretend to be interested and give her my Stevie smile.

The descent happened so quickly. At first I hid my drug use from her. More out of shame than anything else. I just didn't want her to know, because I didn't want to discuss it, and keeping it from her was easy, because she wasn't looking. She was naive to my ways, so as long as I was careful, it wasn't hard to conceal. Maybe she was also looking the other way a little. We both loved each other a lot and when you're in that phase of the relationship, you try to avoid any ugly confrontations.

I spiraled downward as the drugs took over and soon I became the selfish prick from hell. I used to look forward to Cheryl coming home. The simple bliss of having dinner together, watching a movie, or just fucking nonstop ceased. I soon started to resent hearing her keys in the door. It made my skin crawl. I resented the fact that I'd have to hide my stuff before she came back. That was classic H addict behavior, but I wasn't willing to cut it off to save the love of my life.

Soon, I resented the fact that I had to have that big smile ready for her. I began to get sloppy, and one day she came home and my shit was all over the blanket. There were burn holes in the quilt, there were pipes on the carpet, and I couldn't have fucking cared less. She pretended not to see it as she leaned over to kiss me. Then she said she was going to take a quick shower, which I took to be my cue to hide everything. I even resented that. Any lingering guilt was banished and the drugs became all that mattered. Steven the selfish motherfucking drug abuser took over and the light went completely out of my life.

And I mean literally. I actually hung blankets up over the windows because the blinds didn't do a good enough job of blocking the light and blacking out the entire living room. Cheryl thought it was funny at first and tried to make a joke of it. But then I would just scowl at her and go to the bathroom for an hour.

It is all so tragic when I look back at it. I snuffed out this sweet creature's love for me because the other jealous lovers in the room, Mr. Brownstone, Mary Jane, and Joe Blow, couldn't stand her. Cheryl learned to stop making suggestions about what we could do for fun. She learned to stop talking to me. She could feel my body recoil at her touch, so she learned to stop holding me.

Anything could set me off and soon Cheryl was spending as much time out of the house as possible. I stopped bathing. I wore the same shirt for two or three weeks. I wasn't thinking or caring. I was totally self-absorbed. This was it, the lowlife's high life.

## ⚔ HOUSE HUNTING ⚒

When Christmas rolled around, I hadn't been in touch with the guys for a while, since we returned from New Zealand. Then I made no effort to see them over the holidays. Dougie sent each of us a large, framed print of the *Appetite for Destruction* artwork, personally signed by the artist, Robert Williams. It was a nice, thoughtful gift that still hangs in my living room today.

Early in 1989, Cheryl finally got her wish. I got off my ass, and she and I went house hunting. Dougie set me up with a real estate agent. She asked me where I wanted to live and I told her, "Studio

City." They had a list of available homes there, and they drove Cheryl and me to them. Cheryl had a list of homes she thought I'd love. But like an ass, I just picked the second house we looked at. I didn't care to check out any others. It was a two-bedroom home up Laurel Canyon, by an elementary school. I quickly had the master bedroom redone with a higher ceiling, but that was all the redecorating I did, because I spent 99 percent of my time holed up in that room.

The same month, Slash bought a house by a country store on Lookout Mountain. We were right down the hill from each other. He was on the Hollywood side; I was on the Studio City side. It was not far at all, so we started hanging out again.

One day he called me and asked, "You got any money?"

I said, "Yeah, I got money. And you got money too, fucker."

He laughed and said, "Well, come on up." He had his hand out before I got in the door. I gave him $300, and he gave me a thin little piece of heroin. I smiled and got out a crumpled piece of tinfoil. At the time, I didn't realize that he was ripping me off. I was just getting $20 worth here and there, so I wasn't sure what the right amount should be. Meanwhile, he had this huge chunk of the shit. He was shooting it up, I was smoking it.

Somehow I had it in my head that not shooting it gave me some moral high ground to shake my head and feel that Slash was out of control with the shit. Even though I had dabbled with needles, I had backed off a bit and was a little freaked by Slash's behavior. Not long after that first day of scoring together, Slash started to really lose it. We had been partying for a few days, and as the sun was peeking up, I couldn't find Slash in the house.

I went out back, and he was sitting by the pool. He was so out of it, just blindly jabbing a syringe into his arm, over and over. I said "Dude. Stop it. Look, just come in the house. We'll watch a movie, and after that if you want to party some more, we will. Just stop it for now." He didn't listen. He just gazed at me, but at least he had stopped the pincushion routine. About a half hour later, I was sitting in his living room watching TV when he came in, walked right past me, and went upstairs. The last thing I heard was his bedroom door slamming shut.

## ⚞ IZZY, SLASH, ME, AND MR. BROWNSTONE ⚟

Another night, Slash and I paid a visit to Izzy at his new place. He had a loft in his apartment where he would hide from the world, shooting smack and smoking coke. We came by unannounced and evidently disturbed him. He was all weird and strung out from the drugs. He just said, "Hey," and kind of circled the room a few times, scratching his shoulders and his head like he had lice or something. A few minutes later he headed up to his loft. He was completely strung out, yet at the time we thought little of it.

Slash and I were just hanging out, assuming Izzy would show when he had done whatever. Eventually, through our moronic haze, we realized that Izzy didn't want any company, so we let ourselves out. We hit up the Rainbow and Barney's. It was so pointless; we were zombies. At Barney's, Slash nodded out with his mouth full of chili. It started to drip out of his mouth and go all over his shirt. When I got back from the bar, I shook him awake. We ambled out of there and headed back up Laurel.

## ⚞ REHAB THIS ⚟

During the last week in January 1989, Dougie suggested I go into rehab. I remember his remark didn't make me angry, take me by surprise, or hurt my feelings. But like all true drug addicts, I didn't think I was having any trouble with drugs. Dougie sat me down and, without being preachy or pushy, convinced me that it might be a good idea. I trusted him, and I guess I felt deep down that I could use a little tidying up. That was proven by the fact that I didn't make a big fuss about it. This would be my first time in rehab, and I wasn't at all sure what was involved, but I was curious, I guess, and kind of resigned myself to just going with it.

WTF!

When I got out, someone asked me why I hadn't appeared on the American Music Awards. I didn't know what the hell he was talking about. He proceeded to tell me that GNR performed "Patience" during the American Music Awards at the Shrine Auditorium with someone else on drums. I found out later that it was Don Henley of the Eagles who took my spot. I was completely blindsided by this, so stunned and hurt, I can't begin to describe the feeling of betrayal. Nobody in our organization ever mentioned anything about the AMAs to me. My first thought was: "And I didn't even need fucking rehab!"

To be fair, the downward spiral had already gathered momentum and I'm partly to blame for what was certainly the beginning of the end with me and the band. Like the broken finger incident, I'm sure I'd done my fair share to irritate them. I ticked them off just enough to have made them feel indifferent about insisting I be kept in the loop.

Compounding this was the fact that I never let people know if I was hurt or felt cheated. Looking back, if I had been more vocal, I could have at least gotten a feel for what the boys were thinking, and if they were furious with me about something, at least we could have gotten it out in the open. I should have taken the temp and realized trouble was brewing big-time. I kept things so internalized that I never bothered to find out if Dougie made the rehab suggestion just to get me out of the way and substitute Henley. Frankly, I'll never know.

At the time, I was more riled over the fact that they performed "Patience" with *any* drummer. Like the timeless "Lady Jane," "Going to California," and "Yesterday," the song didn't *need* any fucking percussion at all. They could have been noble and told people they chose "Patience" because their drummer was in rehab and they wanted a song that didn't require me to play.

So for all you fuckers who conspired to do me in while pretending to be a close trusted friend, I forgive you. Like I noted at the beginning of this story, God gave me incredible good fortune and I was the one who screwed it up. It's on me, and now I can handle that fact, think back, and just shake my head.

On Valentine's Day 1989, we shot the video for "Patience." It was at an old, abandoned historical landmark, the Ambassador Hotel. This was where Bobby Kennedy was assassinated in 1968 while running for president. I was smoking weed in the kitchen, walking around, tripping out on the tragedy that happened within these very walls. I had the same feeling of loss and doom when I was in Germany, thinking about the Holocaust. It just welled up inside me.

The place was big, dark, and empty. It definitely had a haunted vibe. Very quiet, no one around. I explored the entire building. One of the building security guards asked me if I wanted to see the exact spot where RFK was shot, and I just kind of stared at him. He took me over to this dark corner of the kitchen and showed me where someone had scratched a rough "X" in the dark red cement flooring. I bent down to look at it for like ten minutes. To think this father of eleven children, who had worked so hard to capture the hopes of the youth of America, had come to such a tragic end made me positively ill. I was too young to remember the actual assassination, but I do remember my mom crying in the kitchen when it happened. With Martin Luther King Jr.'s death the same year, this must have been one of the darkest chapters in American history. Everybody loved what RFK and King stood for, and those incredible men paid for their beliefs with their lives.

I pulled myself away from that wretched spot and went upstairs to where they were setting everything up. In the video, we were told that each guy would be given the opportunity to conceive of a silly fantasy short to feature himself. Zeppelin had actually done something like this in their film for "The Song Remains the Same." Duff carried a tray and smoked a cigarette. I wasn't sure why he picked that for his scene. Slash was in bed with a big snake and a hot chick. I imagined a short where I would be sitting between two annoying bitches who were gabbing it up. When the camera panned over to me, I'm just like, "Get me outta here."

We shot the performance scenes back at the Record Plant, where we'd actually recorded the song. Mike Clink was featured in the video working the mixing board. They had the studio done up like a hippie den, with beads and lavish rugs. I had incense burning. It was great, but I couldn't help feeling a cold wind blowing in my direction. Shit, the guys barely talked to me. Maybe I was being overly sensitive, but I was pretty sure that an "us versus Stevie" vibe was starting to fester.

## ⚜ WINDY CITY TIME ⚜

In March, I received word that a state-of-the-art studio was booked for us in Chicago. It would be ours exclusively for two months. Before we left, I asked a buddy if he could arrange a hookup for me when I arrived. He had connections in Chicago. He provided me with a guy's number and I gave him a call. Slash, Duff, and I were the first to arrive. We were under the impression that Izzy and Axl would show up shortly after. We were put up in two condos, which were on the second and third floors of the building.

It was amazing; the very moment we arrived at the condo I was approached by my "new best friend," the drug dealer my buddy had put me in touch with. I grabbed an ounce off him as we entered our new temporary home. All together, we had four bedrooms for us and the crew: Tom, Mike, and Adam. We had various ideas for songs, great songs that Duff, Slash, and I were excited about working on, and the writing for our next album proceeded right away.

The studio was right down the street from the condos. It was awesome. It had a top-of-the-line PA and a grand piano, and my drums were miked. It was located on the fourth floor of a high-rise building. In the basement of the complex was a popular local nightclub. Our presence was kept secret so fans and the press wouldn't mob us. Also, tight security was there 24/7. At night, Duff, Slash, and I would go downstairs to the nightclub, where we would pick up girls and fuck them right in the club. We rarely took them back to the condo.

We'd always have blow on us at the studio. But when I'd offer to cut them a line they would refuse. Then Slash and Duff would go in some other room to party. "Hey, where ya going?" I would begin to follow them only to find that they had shut the door on me. To this day I have no idea why, other than I felt they believed I just wasn't cool enough to hang out with anymore. I ended up doing my blow all alone while they partied together.

Instead of confronting them and flushing out whatever the hell it was that seemed to be getting worse, I let the drugs take me into a dark valley of despair, where I could wallow in my own self-pity. My self-esteem was at an all-time low, and the drugs became my only friend because they would numb the hell out of me and keep the fear and depression at bay.

At rehearsals, I felt I was getting pushed out of the songwriting circle as well. We would be working on the dynamics of a song and the three of us would throw around ideas. Then suddenly the exchange would be limited to Duff and Slash. I learned just to sit and wait patiently. They would agree on something, then turn to me and say, "Okay, Steven, *this* is what we're going to do."

Was I so fucked up that I didn't realize my drum playing was beginning to suffer? Was I lucid enough to even ask myself that question at the time? I believed I was trying as hard as I could to contribute to the songs. But had Slash and Duff sensed that, would they have let me back in the creative mix, or was I already doomed? I don't fucking know. I just don't fucking know. And I think that kind of answers the question. Maybe they felt I didn't care enough to do a course correction. I'm trying to be as honest as possible here, but my emotions were turned so inside out that I find it difficult to look back and figure out what was actually happening.

All I know is that my opinion didn't matter anymore. It bummed me out. We were always a team; it had always been a combined effort. But not any longer, and it was having an affect on the music. I don't care what anybody says, no GNR album was as good as our first one. In Chicago, there was a conscious effort to top *Appetite*. But I believed we needed to take ourselves a lot less seriously and concentrate on the joy and spontaneity that inspired us and had worked so magically on *Appetite*. At times we achieved that, but it seemed much harder this time around. We worked our asses off.

Over the next two months, we wrote thirty-three songs, done. We rehearsed and polished some older songs we had, like "You Could Be Mine," "Don't Cry," and "Back Off Bitch." We totally brought them up to speed. We were proud of these songs. They represented not just the fact that we were more committed, but also the fact that our song craft was evolving.

Seven weeks and five days later, Axl finally arrived. We had two days left in the studio and were anxious to show him all of the new material. He sat there like we were putting him through some kind of torture. Plain and simple, Axl wasn't interested in our material! He just wanted to record a new song he had been working on called "November Rain." He sat at the grand piano in the studio and played it for us. I thought to myself, "That's nice, but that's it?" He had only like two verses written. Duff, Slash, and I had thirty-three songs in the can, ready to go, but Axl wouldn't give them the time of day.

As for Izzy, he never showed.

After Axl settled in, we took him downstairs to the nightclub. By this time hot chicks knew to show up there in hopes of hooking up with one of us in the band. Axl picked up this cute young girl and brought her to the condo. He did his thing where he sits up all night chatting with the girl. But this time his talk, talk, talk, wear-her-down strategy didn't work. She resisted all his advances.

I went upstairs to talk with Tom, and all of a sudden I heard a loud noise and saw this girl come flying out of the bedroom. Unbelievable. Axl was right behind her looking completely crazed.

Duff, Izzy, and I just stood there in shock. She was hysterical. She got up and ran out.

I looked out the window and I could see her running down the street, getting away as fast as she could. I turned to Axl and said, "This time you're going down, dude. You are going to get slammed with some fucking shit for this!"

He just said, "Fuck you," and retreated to his bedroom.

Nothing ever came of it; I guess the girl never said a thing.

## ⊰ BACK IN L.A. ⊱

When we returned to L.A., we continued rehearsing exclusively at Bob Mates Studios in North Hollywood. It was during this time that we wrote a song we would eventually title "Civil War." It's amazing it was ever completed because on most days, when I would come to rehearsal, Slash and Duff would show up drunk. I would get pissed as hell at the guys. I understood that partying went with rock 'n' roll, but we had a record to do. I never put any of the guys down though. Eventually, Slash and Duff just wouldn't show up at all. They may have been hurt or intimidated by the fact that Axl only wanted to work on his stuff.

At this point, Dougie was the only guy I felt any kinship with. The band didn't feel like the band, with so many fucking dramatic undercurrents, whispered gatherings behind closed doors, and members breaking off into pairs. It was no longer GNR, it was some cheap daytime soap. I felt completely alienated from the band.

## ⊰ DOGGIE DUTY ⊱

Months passed and the times I got together with the guys became few and far between. The one time I saw Duff, I invited him to come with me to buy a new dog. I was always fond of an Irish pug owned by my old buddy Matt Cassel. I always said, "When I get my own place I'm going to get a pug." So Duff and I drove out to the Galleria mall. We went to the pet store there and looked at a few pups. We had them take this cute little pug puppy out, and we set him down. He was just a little fur ball darting back and forth across the floor. He rolled over to Duff's foot and pissed on his boot.

"Hey, buster, what's up?" he said. It was such a cute moment, I just fell in love with that dog, and the name "Buster" stuck. I bought him and a few weeks later, I bought another pug, Benson, to keep him company, and together they kept me company.

## ⇥ THE DICE MAN COMETH ⇤

Duff married his Lame Flames girlfriend, Mandy Brix. He had a bachelor party at the studio. Andrew Dice Clay was hired for entertainment. He was the biggest comedian in the world at the time. He was the only comedian to ever play massive venues usually reserved for pro sporting events and major rock shows, like Madison Square Garden, and pack the house.

But that was the Dice Man in his prime. And here he was, a good friend of the band willing to do his hilarious act for this small private event. Ronnie had scored some acid. About an hour before the ceremony, we dropped. During Andrew's stand-up routine we were tripping our balls off, laughing out loud at the top of our lungs at everything he said.

## ⇥ MTV CHARITY EVENT ⇤

MTV contacted our management in hopes of getting some of the guys to participate in the 1989 Rock 'N' Jock softball game. This was an annual fund-raiser that pitted celebrities, TV stars, rock stars, and actual professional baseball players against one another in a televised game of softball for charity. I was the only one from the band who went.

I brought Jamie, my little brother, with me and we had an epic time. When we arrived, we went right into the locker room, where I saw Sam Kinison sitting in what looked like a barber's chair, having some touch-ups applied by a makeup person. I walked over and smiled. "Hey, Sammy," I said. He had his eyes closed and he was mumbling something under his breath, not at all coherent; in fact, he looked like he was nodding out. "How the hell is *he* gonna play?" I thought.

We went into our dugout, where we met the other guys on our team. Tone-Loc, a rap star who had a couple of huge hits at the time with "Wild Thing" and "Funky Cold Medina," was very friendly. He lit up a joint and we smoked out right there in the dugout. He had the best shit. Just before the game started, Sam Kinison came running out screaming at the top of his lungs. He was a fireball of energy, and I figured he must have done some good shit, because only moments before, the guy had looked sedated, completely out of it.

The MTV cameramen were all over the place, interviewing the celebs. I did a little interview and they asked me the question everyone had been asking, "When's the new album coming out?" I told them that we were still working on it and to expect it to be heavier and better than *Appetite*. Right.

The game started, and in the first inning they put me out in left field. I was so stoned, and I began to trip out, thinking, "God, please don't hit it to me, please." The very first pitch of the game, the batter connects, a towering shot right over to me. "Fucking great," I thought. As it sailed over my head, I jumped up, missed the ball, and fell right on my ass. Before I could even get up, the center fielder was already there. He threw the ball to second base and nailed the runner for an out. After three outs, we all ran back to the dugout and I was like, "Fuck this." I sat down in the dugout with no intention of playing again.

Tone owned a barbecue restaurant, and he sent two of his friends to pick up some food. They came back with tons of ribs, coleslaw, and beans, and it was a full-on pig-out. Eddie Money was also there with us. He was pretty huge in the seventies and was riding high on the success of a more recent hit, "Take Me Home Tonight." We were talking about music and, of course, drugs. He rolled up his pant leg and showed me these massive scars he got from shooting up. It was disgusting, and I couldn't believe it. Eddie Money; it just blew my mind. Are hard drugs truly the price of admission for rock stars? Here was just another example.

# 17

# MARRIAGE *and* DIVORCE

## ⊰ DOWN THE AISLE ⊱

After the softball game my natural defenses kicked in again and I backed off the hard stuff for a while to regain a certain degree of clarity. In fact, I soon felt sharp enough and smart enough to plot a way back to feeling 100 percent better. I realized Cheryl had been amazing through the whole ordeal of putting up with me while I was using heavily, and through it all she never abandoned me. She never nagged me, she let me do my thing, and whether it was making lunch or making love she was totally there for me.

So I asked Cheryl to marry me. When I proposed, she couldn't have been more thrilled. I booked a flight to Vegas, and we decided to fly out and get hitched. Just like that. I called Dougie and told him the great news. "Oh, no, you're not," he said. "Listen, Stevie, you don't know what the hell you're getting yourself into."

I didn't care what he said. I told him, "I love her, Dougie."

Doug shot back, "Well, wait just a few hours. Please. I've got to bring some papers over for you to sign anyway." That afternoon, he brought over a prenuptial agreement that Cheryl had absolutely no problem with. I knew that would be her reaction. We were in love, and she was the most sincere, honest girl I had ever known.

We arrived in Vegas and got married the same day. No bachelor party, no bridesmaids or ushers, no reception, just Cheryl and me down the aisle. I remember looking at the marriage certificate and was amused by the date. Totally randomly we wed on "6/7/89." How sweet is that? Even I can remember my anniversary.

## ⇥ A CLOSER FAMILY ⇤

A week later we returned home and I received a call from my mother. She was beyond upset. "Steven. Why didn't you tell me? I was in line at the grocery store and there on the cover of the *National Enquirer* was the news that my son had married!" She read the headline to me: "Guns N' Roses drummer weds. Wife signs agreement allowing him to cheat." I thought that was so funny, I made the mistake of laughing out loud into the phone. Mom didn't see the humor.

Of course the headline wasn't true. "This was not the way to find out about the things you're doing, Steven. I want us to be a loving family. Your getting married should have been one of the proudest moments in my life. Instead it's brought me pain and humiliation." I honestly didn't feel that bad about what I had done. But deep down it must have bothered me because I brought Cheryl over to meet Mom and Dad that same evening. It was strained company at first, but after a few toasts, and Cheryl's loving manner with Mel and Ma, things got very nice. I gave Mom a big long hug before leaving, and to my surprise, I was the one with the misty eyes.

After that, I made a conscious effort to be more in touch with my family. I would even pick my little brother up from school from time to time, something Jamie really loved. I would drive my Mercedes to school, or the new black Ford Bronco that I had just gotten. I bought it from Andrew Ridgely, who was famous for being in Wham!, the band he shared with George Michael.

I'd tell Jamie I'd be picking him up at the parking lot by the school football field. Lots of kids would be waiting for me there, and at times it seemed like the entire student body had turned out. I'd hang out and sign autographs for everyone. Finally I'd say, "Okay, bro, we gotta get going." He'd hop in, and we'd take off. The smile on Jamie's face said it all. I could occasionally be a great brother; I just couldn't always be a good brother.

On his sixteenth birthday, I took him to buy a car. I said, "Get whatever you want." I recommended a truck to him, but he ultimately decided on a brand-new Chevy Camaro Z28, all tricked out with a great sound system, special rims, leather interior, and the "racing package," which added about a hundred horsepower to an already powerful engine. Hey, it felt great just to see the look in his eyes when they rolled it out. Looking back, I realize that as the band and I grew more distant, my family became more important to me.

In September 1989, Dougie called to tell me that the band would be opening for the Rolling Stones at the Los Angeles Coliseum next month. I was so stoked. Maybe my fears were unfounded, because all my dreams were still coming true.

We were to do five shows with the Stones in late September and then go back to a place called Mates Rehearsal in North Hollywood to rehearse for the *Use Your Illusion* tracks. I felt wonderful after hanging up with Dougie. Everything was going great again. And maybe all this concern about my being marginalized by the band was just baseless worry.

During this time, Living Colour was growing in popularity as a black metal/rock act with a hit called "Cult of Personality." Their guitarist, Vernon Reid, was an outspoken black activist and publicly took offense to the lyrics in "One in a Million." His music career must have trumped his personal beliefs, because Living Colour agreed to open for *us* during the Stones shows.

Axl had a limo pick him up from home and take him to the shows. Slash, Duff, Izzy, and I were put up across from the Coliseum. Cheryl and I stayed there, and I would walk over to Slash's room to hang out and party. Unfortunately, every dealer on the

West Coast was buzzing around for the concert, and I fell to temptation again. At this point, Slash hadn't let up at all and was getting sucked deeper into hard drugs. Heroin came packaged in rubber balloons, and that night after we checked in, I bought six of those balloons and went to Slash's room. I walked in and I saw Slash in the bathroom, and he had like twenty of these same balloons lying around, already opened and used. He was just sitting on the toilet, staring down at the tiles, all stoned out. He was going to be no fun, so I just spun around and left.

## THE GREATEST ROCK BAND IN THE WORLD

We got to meet our heroes the first night before our performance. I was surprised by Mick Jagger's appearance. I thought he was a little skinny guy from all those videos, but when he walked in the room, he had the presence of a giant, and he was in great shape, buffer than buff. I mean, he was cut. *Life* magazine once ran an article about Mick prepping for Stones tours, how he would get on a strict diet, run every morning, and lift weights like a boxer prepping for fight night. It looked like he was still devoted to that routine.

The whole band was there but Slash, who missed out because he was getting high. In fact, he just made it to the stage for our show. We were all partying pretty hard those days. As I neared the stage, I could hear the fans. As I rounded the corner, I could see the multitudes screaming their heads off.

The sound of that crowd was so powerful that it actually gave me an incredible buzz. When the audience caught sight of us, they all bolted upright. It was like one giant wave of energy, intensely stimulating. We were the proud prodigy, the bastard sons of the Rolling Stones, and we killed that night. We were there to show the world that rock was alive and bigger than ever, and we succeeded in every way.

**B**ut at a time when we should have been rejoicing beyond all measure, Axl instead chose to wag his finger. He had become aware of the out-of-control partying that was happening within the band and he made a long rambling statement during the second show. "If some people in this organization don't get their shit together and stop dancing with Mr. Brownstone, this is going to be the last Guns N' Roses show. Ever!"

Axl went on and on, threatening to shut us down if the runaway abuse continued. Maybe it was done for publicity, maybe out of genuine concern, I don't know, but it was way over the top. Disbanding GNR for drug abuse was like grounding a bird for flying.

So we all had to snicker when the Stones took the stage and Jagger decided to bust Axl's balls for his little lecture. He stood up there, smiled, and grabbed the mike like he owned the whole fucking world. He strutted to the very front of the platform, leaned out over everyone, and waved his arm, asking the crowd if they had "heard enough of Axl's bullshit" and were ready to rock 'n' roll. Of course the crowd's response was a deafening affirmative.

Axl's statements made national entertainment news the following day, and no one said a goddamn thing about it. I had learned my lesson, so I wasn't about to be the one to start. But sadly, no one else did either.

For the most part, Axl had been ignoring me during this period. But that was my fault too. I never took the initiative to talk with him and find out what was simmering in that brain pan of his. I wish I had insisted on making the time to sit him down and sort things out to clear the air.

In addition to our rooms across the street, each of us was given our own trailer on the Coliseum backstage lot to hang out in before the show. MTV was making a rockumentary about us and visited each of us in our personal trailers for interviews. I was hanging out with Cheryl, Ronnie, and David Lee Roth. David Lee was just being DLR, the legendary front man and incredibly funny friend.

My family was extremely excited about the event so I made sure to have Dougie take care of them. He sent limos for them every night. I saw them only briefly, however, because when I was performing, particularly in something this momentous, I was in my own separate world.

On the night of the last show, a unique thing happened. At the end of our set we put our arms around one another, and as a group, we took a bow. We had never done that before. It felt kind of awkward but appropriate. In my mind, that show was the last real Guns N' Roses concert ever. Immediately following that bow, we once again went our own separate ego-inflated ways.

## ⤜ KNIFE IN THE BACK ⤏

In early 1990 the band agreed to appear at a benefit at the famous Hoosier Dome in Indianapolis called Farm Aid. It was huge, tens of thousands of fans cheering nonstop, with millions more watching on TV. While it was an important event, we didn't even bother to rehearse for it. I flew out there expecting to have a great time, but Duff and Slash continued to distance themselves from me. They seemed locked into their private little clique. Izzy was off on his own, but that was typical.

So I found myself hanging out exclusively with Dougie. No one else was talking to me. I felt very isolated. After that Stones show everyone kind of withdrew from me again, and the excitement I had felt during the event evaporated.

When we were introduced at the Farm Aid concert, I was so excited that I sprinted out to the drums, and as I leaped up, I caught my foot on the flange that ran around the border of the riser. I tripped and fell right on my ass. I might have been a little buzzed, but let me tell you, there's nothing like wiping out in front of all those fans to sober your ass right up. I was bummed—"Shit, I'm on live TV." But I quickly scrambled right back up, smiled broadly, and grabbed my sticks, ready to rock. I assumed we'd be playing a couple of our hits, like "Paradise City" or "Welcome to the Jungle." Axl announced, "This is something new we got, called 'Civil War.'"

Huh? Although I knew the song, I didn't know that would be the title. So I looked at Duff and I was like, "Dude. What's goin' on?" He was kind of being a dick, maybe disgusted with my wipeout on the stage, so I just sat there, and when I heard Slash play the opening riff, I caught on. Although we didn't even have that song completely down and had never rehearsed it with Axl, it played pretty well. I kind of sighed with relief to have gotten over that hurdle, but the damn surprises kept coming.

Next Axl says, "This is by a punk band called the UK Subs. And this song really rocks; it's called 'Down on the Farm.'" I'm like, "What the fuck?" I yell over to Duff, "Dude! How does it go?" He just claps his hands, providing me with a tempo, and then walks away. So I just played the tempo with my bass drum and winged it. I'd never once heard that song before. But I kicked ass, and that made me feel proud, not mad.

Looking back, I realize that this may have been proof positive that their plan to get me out of the band was already in full motion. They weren't cluing me in to new songs or even telling me what they were playing. I believe their strategy was to make my playing sound like shit. I believe they wanted me to fuck up on live TV; that would be their evidence. By branding me as an ill-prepared, crappy drummer, they'd be armed with a sound reason for kicking me out.

## ⊰ GETTING OFF THE STUFF ⊱

When we came back to L.A., we again went our own ways. I had gotten to another one of those junctions where my body was warning me to stop partying. I hit the brakes for about a week, then I suddenly became very ill. I had no idea what was wrong with me. I had been smoking heroin regularly and I was giving it an indefinite break. Now, I was shaking all over, feeling very hollow and cold. I was experiencing the full-on blunt force of withdrawal, as my body ached like it never had before. I lived in the bathroom, constantly having to throw up.

I called Dougie and told him what I was going through. He told me he wanted to take me to the doctor right away, and I immediately calmed down, thinking, "Good ol' Dougie, looking out for me." So we went to a medical facility at Olympic and Fairfax. The doctor there broke off about a quarter of a small pill and had me take it with water. He explained that it was an opiate blocker and told me, "This will make you feel better, because even if you try to cheat and take heroin, you won't feel a thing." What they didn't tell me (and what the fucking MD didn't bother to check out first) was that you needed to be completely clean to take it. Patients needed to *detox fully* in order for the drug to work properly. If you had opiates in your system when you took it, it would *fuck you up*. God, did I discover that the hard way.

Within hours after returning home, I became deathly ill, even worse than before. I called Dougie and told him, "Whatever the fuck they gave me isn't working. I'm sicker than I've ever been in my life!" He sent a registered nurse over who was qualified to examine me. After she left, I remember sitting down, momentarily relieved that I'd be okay now. But as the sweat began to pour down over my face, I suddenly became incredibly scared and honestly thought I was going to die. This feeling lasted an eternity, because as I said, I hadn't completely detoxed. You'd think they'd ask you your status before giving you pills and injections. I was terribly sick for weeks. Then came the deathblow: Slash called me and told me that we were going into the studio to record "Civil War."

"Dude, haven't you talked to Dougie? I'm sick as hell."

Slash didn't want to hear it. His voice was strangely detached, zero emotion. "We can't waste any more money," he replied.

Was I really hearing this shit? From my dearest friend, the guy I was instrumental in getting into GNR, for fuck's sake? Where was the loyalty, the compassion? "Fuck that, Slash. Listen to me. We both know someone in the band who's wasted a helluva lot more time and money than it would cost to postpone this one lousy recording session. It would just be for the week or so that it would take for me to get better." We hadn't done shit in over a year and now they wanted to record one damn song, and they couldn't wait for me to feel better. It was such bullshit, and I could only hope that it was someone else pushing their buttons. I didn't want to believe that Slash really had it in for me.

## ✄ DUMPED FROM GNR ✄

With no alternative, I attempted to do my job. I literally pulled my head out of a toilet, showered up, and got to the studio on time. I sat on the stool, staring at my drums, but another wave of nausea hit me and I was suddenly sick as hell, doubled over in pain. The guys looked at me, and there was no mercy in their faces. Nothing.

Instead, they were annoyed with me, and no one said a thing. I tried to play but my timing was off. The guys in the sound booth asked for take after take, and finally I couldn't take the tension. "Guys, I'm fucked up. But I'm sick, not high. I'm just ill and that's all." I asked Dougie to clear the matter up for me. "Dougie, tell them. Tell them how sick this medication is making me."

But like a waking nightmare, Dougie looked away. I pleaded with him: "You've got to tell them that even if I *was* partying, the medicine they're making me take would block it." Dougie didn't say a word. My last buddy abandoned me. There was no love; he just turned and left the room. I had been set up, through my own stupid actions, and they wanted the absolute worst for me.

I never thought this could happen to me. It was always the five of us united, an inseparable team. But the Guns N' Roses machine had become massive, and I could feel it shoving me aside. I couldn't stand the idea of being pushed out of the band. I desperately didn't want this to end, and I honestly thought I had done nothing to deserve having it taken away from me. I just did what we all were doing, living the rock star life.

I seemed to be suffering under an unfair double standard. Christ, we open for the Rolling Stones, and Axl falls off the fucking stage while singing "Out ta Get Me." The whole thing's treated like no big deal. But *I* misjudge the drum riser during Farm Aid and the response is total outrage; "Look at Stevie, that drugged-out waste of an irresponsible fuckup." We had all worked so hard to get to the mountaintop and were just beginning to reap the rewards. In my worst nightmares, I never imagined that it could all be taken away from me.

I counted on Dougie to keep me in the loop. He had me believe that he had my back, that he cared for and loved me. Well, he fooled the hell out of me. I had been lured into having total trust in him and didn't want to believe some conspiracy was actually going down.

The day after the "Civil War" recording session, Doug called me and asked me to come down to the office to sign some papers. He offered no explanation for his behavior the previous day, and I didn't try to lay on any guilt. I just told him I was still very ill. There was a long silence on the phone, then Dougie told me that the matter was very important and wouldn't take long. He told me he had been instructed by the GNR attorneys to tell me that my presence was absolutely required. In spite of what had gone down, I still wanted to believe that Dougie was my caring wingman, and when he promised I would be in and out of there quickly, I decided to rally. I cared more for his situation than my own. I could hear the stress in Doug's voice and I didn't want to bust his balls, so I got myself together and Cheryl drove me. When I walked in, Dougie and one of our lawyers, a professional-looking middle-aged woman, had a stack of papers for me to read.

Read!? I couldn't even see. They told me all I had to do was sign at the bottom of all the pages with the colored paper clips attached. I asked what this was all about. Dougie told me, "It's nothing to worry about." In my condition, I wasn't about to read all this shit, but I was a little freaked and my jaw just dropped. In essence, I thought I was agreeing not to party and not to screw up on any band-related activities for the next four weeks. If I fucked up, they would fine me $2,000. I thought, "What the hell, no problem. The band doesn't even have anything scheduled during the next month, and even so, what's two grand?" I signed everything. I just wanted to get out of there, go home, and lie down.

I discovered later that what I had actually signed away was *my life*. What the legal papers actually stated was that they were going to give me $2,000 for my contribution to Guns N' Roses. Everything else, my royalties, my partnership in the band, my rights, was gone! Of course, I didn't know this at the time. I'm sure with all these papers I naively signed, they thought they had my fate sealed. They had a signed, ironclad deal against me.

The next afternoon, I received another call from Doug. "The guys don't want you to be on the next record. They are going to use someone else."

I was still feeling like shit, and at this point I guess I saw it coming. "Yeah, whatever." I just hung up the phone and started crying. I'd had enough, but I couldn't help but be depressed. I didn't even bother calling Slash. What was the point?

To blunt the pain, I went on a party binge, smoking weed, drinking Jägermeister, and popping whatever pills I could find. Cheryl was there with me, and she would never say anything to upset me. She was there by my side, but I didn't care and wasn't even aware of her. I just locked myself away in my room.

Cheryl didn't fully understand what was happening. And with all this heavy shit going down around us, I couldn't handle it, wouldn't handle it. She wasn't prepared to deal with all the crap either, and every day she cried a lot because she knew something horrible was occurring.

*And I'm standin' at the crossroads,*
*I believe I'm sinkin' down.*

—"CROSS ROAD BLUES," ROBERT JOHNSON

felt I had sold my soul for rock 'n' roll, and the devil had just stopped by to stamp me "Paid in Full." A couple of days of partying only put me in worse condition, and I came out of my stupor so depressed, I tried to kill myself. I slashed my wrists, suddenly became very light-headed, and collapsed onto the hard floor. My face must have hit a chair or a coffee table as I fell because Cheryl raced in to find me badly bruised, with my lip split wide open. The cuts to my wrists weren't nearly what was required to do the job properly, but they left ugly scars that still remind me of this dark time.

I believe I was crying out for help more than actually trying to die. Cheryl called Doug and told him that I was very fucked up and had tried to kill myself. That afternoon, Doug, Slash, and a security guy named Ron came to my home. When I opened the door and saw them, I panicked for some reason and just took off trying to run away from them in my own house. I know that coke eventually makes you very paranoid, but there was no reason for me to be scared of these guys. In an even dumber move, Ron went chasing after me. I don't know what he was planning to do when he caught me. I hopped out an upstairs window and ran along the roof to the top of the garage. They were yelling up to me: "Steven, come down. Come on, man, come down."

"No. Fuck you, fuck everything!" Then I just dropped onto the roof, crying like a baby.

I heard a noise and realized they were going to climb up and get me. This gave me an even worse panic attack so I jumped off the roof of the garage. I plummeted into the cab of Slash's black truck. Everyone was shocked and just stood there as I bounced, unhurt, then rolled off to the ground, a total mess.

The security guy was a supreme asshole. He dismissed the whole matter like I was a piece of shit, not worth the time. "The hell with him, let's go." It was as if they were looking for any reason to leave, so on Ron's remark, they split.

The next evening, Slash phoned. Inwardly my heart thumped, and I felt like here was my old friend, reaching out. But no, he was actually pissed. "Dude, you dove on my truck, and it's fucking dented. You dented my truck, and you're paying for it."

I was numb. "Whatever. Sure, I'll pay for it. No problem, buddy. Take it out of my two thousand dollars, you heartless piece of shit." But at that point, all I heard was a dial tone.

*Fame puts you there where things are hollow . . .*

—"FAME," DAVID BOWIE

## ⊰ DEATH BY DEGREES ⊱

Looking back, I still cringe at this dark, torturous time in my life. Up to this moment, I had been high practically all the time and that made me careless, among other things. But in all honesty, I was the only member of the band who was held accountable for that carelessness. And now my situation was hopeless. I achieved the dream of a lifetime, and just as it was about to blossom fully, they stomped on it. I was riding high; the group that I had formed with my friends just five years before had become the biggest rock band in the world.

It seemed everyone wanted to know me, and I was very touched by the way I was treated by our fans. Everyone was so affectionate, and I tried to return that love in spades.

I really felt blessed and thanked God for my good fortune. People said, "Enjoy this. Take it in as it's happening. Try to live in the moment." That's all I ever did. It was the way I welcomed each day naturally. I didn't have to remind myself to try to live in the moment because that was simply the way I had always experienced my life.

There's abundant proof of this. Look at the videos of me playing, I'm the only guy in the band smiling, loving every minute of it like no one else. I was constantly aware of God's grace and was thankful for it. I hugged everyone who wanted an autograph, sat and talked with anyone, and freely reached out to the people who approached us. From anyone's perspective, I honestly believe that it's clear I was the one who truly savored our success the most.

When girls would say I was the cutest or the sexiest or the nicest boy in the band, I would just laugh. And I'd always be sure to spread it around, telling them Slash was much sexier, Duff was much nicer, Izzy was much cooler, and Axl was smarter.

Ronnie Schneider and I went out one evening to a club called Bordello. This was just before news of my getting kicked out of GNR was made public. Bordello was a popular hot spot located at Santa Monica and Fairfax. As with any trendy spot, there was a line with dozens of people waiting to get in. We got there and stood in line with everyone else. I noticed the door guy peer over the line in my direction. He walked over to us and said, "Steven Adler. Guns N' Roses! What are you doing here? You don't have to wait in line!" He put his arm around my shoulder, walked us to the front, and opened the door as if we had been buddies for years. I thanked him, shook his hand, and entered the club. Fact is, I hadn't minded waiting in line. I enjoyed talking to everyone but was of course thrilled to get in. Ronnie and I had a great time that night.

By the end of the week, the news hit the world that I was no longer in the band. To add insult to injury, I was portrayed in the news as the consummate loser. "Band that glorifies drug use fires drummer for being out of control on drugs." If that doesn't make me sound like the most pathetic person on earth, I don't know what would.

I felt that familiar chill cut through my heart again, that emotional emptiness that meant my family had abandoned me. And GNR was my family. Izzy, Axl, Duff, and Slash were my brothers; we loved and cared for each other, had each other's back, and fought like hell to succeed together. Now, I was no longer welcome in my own family. Again!

God had given me a second chance and I blew it big-time. I desperately needed to be numb, to just take away the pain. By the end of that week, all I could do was sit in my house smoking coke and heroin.

Eventually, Ronnie, remembering the great time I had at Bordello a few weeks earlier, thought it would be nice to get out of the house and go to a place where I could feel wanted. Again there was a line at the door. Confident, I walked up to the doorman, the same guy, and greeted him enthusiastically. "How are you?" I asked.

He looked at me and seemed annoyed. I stood there for a second. "What do you think you're doing? You gotta stand in line just like everyone else." He pointed toward the end of the line, making a scene for all to see. I was shocked but waved him off and walked away. A block down the road, my emotions got the best of me. I had just been treated like a piece of shit, and that's how I felt. It was harsh. I walked home with Ronnie and continued the assault on my pain.

## ⊱ RAY OF HOPE ⊰

Shortly after, I stopped going out altogether. All I wanted was to be alone and even refused the love of my wife. Cheryl was having difficulty dealing with me and the entire situation as a whole. I feel horrible to this very day; putting her through so much drama was not fair at all. One of us, I think it was probably Cheryl, decided that it would be best for her to take a break for a few days and visit her family.

Just when I couldn't have been more numb or depressed, hope appeared on the horizon. One of my lawyers called and told me that AC/DC was auditioning for a new drummer. "They are considering you, Steven. I am going to get you this gig."

"Do it!" I shouted. I was so happy; at last, a chance at redemption.

But the stars were not lined up for me. That same fucking night, an interview with Axl aired on MTV. He spoke of how GNR was so much more than he ever expected. Then the topic of the former drummer came up, and Axl stuck a spike in my heart. "Steven is so fucked up on drugs. He can't even play anymore. He's someone I used to know." My head was spinning; this was on MTV, national

TV. Axl, the most popular rock star at the time, had just told the world I was a fuckup. It was unbelievably bad timing. I never heard another word about the AC/DC gig.

After a couple of weeks, Cheryl returned, taking a cab from the airport. She yelled and screamed at me when I answered the door: "I tried calling you. You can't answer the goddamn phone? I thought you were dead!"

I could barely mutter, "Sorry, honey." In fact, I hadn't thought of her in days. She could have been gone a week, a month, and I wouldn't have known because time no longer had any meaning. I was beginning to sink even farther downward, carving out a routine that would become my degenerate way of life for a major portion of the next ten years.

## ⊰ THE WRATH OF AXL ⊱

My fate was sealed when an unforeseen run-in with Axl sent my entire existence into a permanent tailspin. Right after Cheryl returned, I found out that Andy McCoy had moved in just down the street from us. Andy was the guitarist for the band Hanoi Rocks. I was introduced to their sound through Axl and Izzy and instantly fell in love with their brand of hard-driving rock. In fact, Guns N' Roses' own label, Uzi Suicide, had just released Hanoi's entire back catalog on CD for the first time. They were the only other band to be released on GNR's label. I was disappointed, however, to learn that Andy had married Laura, Izzy's ex-girlfriend. I hated this woman. She was attractive enough but such a bitch that I considered her repulsive. Desperate to keep the music in my life, Andy and I started to hang out and jam. I hired the guy who remodeled my bedroom to turn the tool shed into a small studio. He soundproofed the walls and really did a great job. It was a bit cramped but it didn't matter. Andy and I worked on new songs and jammed out on some classics. Andy knew I disliked his wife, but it had no effect on our friendship. I just told him I didn't want her around, and that was cool with him. But I guess we were getting along so well that eventually he decided it would be no big deal. One day, while I was in my yard, I could see the two of them walking down the hill, clearly on their way to my place. I just stood there, giving them both two high and mighty middle fingers. Andy caught my gesture, but not Laura. Andy never broke stride and just walked in with her as if it was a nonissue. What could I do? It's no secret I'm a softie at heart.

Before long, Laura was dropping in regularly and getting on my nerves like she had in the past. One day, she pulled the most fucked-up stunt ever. Andy and I were jamming in the shed when we heard a pounding at the door. I opened it, and there was Laura with Axl's fiancée, Erin Everly.

Erin was completely out of it and could barely stand up. I asked, "What the fuck is she on?"

Laura said, "Nothing. She and Axl had a fight. Can you give her something?"

"What? I ain't giving her shit," I yelled, and grabbed Erin, who was swaying back and forth, eyes closed. "Erin, are you okay? You better—"

Laura interrupted. "C'mon, Steven, just give her something."

"What is she on? What did you give her?" I yelled at Laura.

She said, "Steven, she and Axl went at it, so I gave her some Valium."

I screamed, "What the fuck is wrong with you?" Erin could barely stand so I carried her into my bedroom and set her down. "Erin, are you okay?"

Her eyes opened slightly. "Axl and I had a fight."

Laura came in. "Steven, chill. I just gave her a few pills," she repeated, not the least bit concerned.

"How many?" I yelled.

Erin was starting to go out and I panicked. I wasn't taking any chances with this. I called an ambulance and tried not to freak the fuck out. I did not need to be involved in this situation. She was Axl's bride-to-be for chrissakes. The paramedics arrived, checked her vitals, and told me they would have to induce vomiting. Before they whisked her away, they assured me that her pulse was strong and she would probably be okay.

Later I discovered that Erin already had heroin in her system. When questioned, they said that I was the one who had given it to her. Axl called and threatened me: "I'm coming over there and I'm going to fucking kill you!"

I yelled, "I didn't give her shit."

"Bullshit!" he said.

I was livid and screamed back, "I didn't. Fuck you!" I hung up the phone.

My heart raced, and I truly believed that Axl was furious enough to want to kill me. I began to fear for Cheryl's and my well-being, so we rounded up the dogs and took off for Palm Springs. Axl told the press that I shot Erin up, and no one had any reason to believe otherwise. Not that it could do any more damage, but the guys in the band thought I was an even bigger asshole than before.

I would never shoot heroin, or any drug, into Erin. I always adored her, and probably helped to save her life that day, but it didn't mean shit. I couldn't fucking stand it. I was completely miserable and my existence became even more unbearable, if that was possible.

# 18

# HIGH *or* DIE

## ⚞ AFTERMATH ⚟

The Erin debacle cemented the beginning of a dark and destructive period where I quietly burned in my own private hell. After getting kicked out of GNR, all I cared about was getting high, and if that meant dying, so be it. Or at least that's what I was telling myself. God, I was scared. I couldn't even admit this to myself, but deep down inside, I probably didn't trust myself to be alone. With each day, Cheryl witnessed my becoming more despondent and withdrawn. She wasn't around for much longer, however, and I couldn't really blame her. One evening, while I was lying in front of the fireplace, high out of my mind, I heard a faint voice: "Stevie, Cheryl's gonna stay with me for a while."

I opened my eyes to see the blurred image of one of Cheryl's girlfriends. I managed to mumble, "Okay."

It would be years before I saw Cheryl again.

This abandonment was devastating and I went from bad to worse, becoming even more destructive. Cheryl was my last bit of support, and she was gone. It was my own fault because I had ignored Cheryl completely. She honestly, desperately tried to get me to help myself, but I was too far out of it. I was way beyond asking for anyone's help; in fact, at this point, I just wanted to fucking die. "Dear God, just let me die."

Every now and then I'd experience a moment of clarity. Guns N' Roses had a great guy named Todd on their road crew. He went on to become Skid Row's tour manager. When I became really sick, he let me stay at his place in Arizona. He had a speedboat that I would take out on the lake. He also had those ATVs, great for tooling around. Since I didn't know anyone in Arizona, it was impossible for me to score dope.

I also went to Hawaii, but it was incredibly hot and humid. The big convenience chain store there is called ABC. There was one right across the street from my hotel. It was so scorching that after walking to ABC and back to get smokes, I'd have to take a shower. It was that miserable. I usually took friends with me to Hawaii. Ronnie Schneider and a close mutual friend of ours, Steve Sprite, came along a few times. Sometimes I'd leave for a week, or other times just for the weekend. It felt pleasant to dry out and get a decent, natural sleep cycle going. But my dry spells never lasted long. I'd get sick of the weather and as soon as I returned home, I'd make a call and the self-destruction would kick in again.

On my way to score dope one afternoon, I was cranking the stereo in my Mercedes. When I was on tour with GNR, someone had given me a tape of an up-and-coming band from San Francisco called Vain. I fell in love with their straight-ahead rock 'n' roll sound. I was cruising on Santa Monica Boulevard listening to one of their albums when it hit me. "Hey, these guys aren't doing nothing," I thought. A lightbulb lit up and the excitement meter dove into the red. "I'm gonna get in touch with these guys and we are going to create a kick-ass band."

I had already met the guys in Vain a couple of times in the past. When GNR hit San Francisco they would come out to the shows. I made a few calls and before I knew it, I was in touch with their front man, songwriter Davy Vain. He shared my enthusiasm for starting a new project. Davy got in touch with a couple of his former

Vain bandmates and made a proposal on my behalf. Their guitarist, Jamie Scott, was working in a music store, a gig he had no problem quitting. Guitarist Shawn Rorie and bass player Ashley Mitchell fell into the fold. I flew them out to L.A. and rented a studio, and we proceeded to see if this thing would work.

I already knew all the Vain songs from the tapes, so we practically had an entire set ready right out of the gate. The chemistry was great, and I thought this definitely could be the new kick-ass band I so desperately needed. Davy knew some experienced businesspeople and secured the services of a lawyer and a manager. They were responsible for getting us gigs and press.

I had Davy move into my guesthouse, and I put the rest of the guys up in two apartments just down the street. In all, it cost me a couple thou a week. I never thought too much about my finances; I had people doing that for me. I just felt that this was the one thing in the world that was worth every penny.

Joining the guys from Frisco was their roadie, a frighteningly big guy named Rocko. He had crooked yellow teeth and stringy red hair and was the epitome of a big, ugly redneck. I felt a little uneasy about him at first, but I figured, "Shit, Jamie wouldn't bring someone on board if he wasn't cool." Rocko knew his shit professionally and was an exceptional roadie. Like everyone else in my life, once I had made up my mind, I accepted him with no problem and greeted him with a smile.

We started to rehearse every day. We performed the entire Vain catalog and worked on a couple of cover tunes, most notably a rocked-out version of the Jimi Hendrix classic "Voodoo Chile." Unfortunately the commitment to solidifying this bitching new band wasn't enough to keep me fully occupied. I still partied regularly, and the guys caught on to my wicked ways pretty quickly. Sometimes I'd miss rehearsal because I was waiting on a dealer, or I'd be too fucked up to even play. The guys just wanted to rehearse, play it loud, and get it tight. They didn't really party at all. They would drink once in a while, and that was about it.

But for me, there was still enough pain there, or just plain bad habits, that no matter how excited I got about the music, it was only a temporary stay from the drugs. Using was still front and center; the music just became a healthy distraction.

## ⊰ PLAYING IN NEW YORK ⊱

In spite of my continuing to party, we rehearsed and recorded an eight-song demo, which I believed completely rocked. We did it in the same studio where Metallica had just finished their "black" record. I know this sounds stupid and irresponsible, but we didn't zero in on an official name for the band until we finished the demo. I'm sure it had been swimming around in the back of my head because as soon as I thought about it, I recognized it as a name I loved and had even used on my first car, the Road Crew-zer. I should have legally secured it as soon as I realized we couldn't imagine calling it anything else. But we waited a bit too long, and by that point, Slash had learned of my new Road Crew incarnation and sought registration before we took the necessary steps to protect it. I couldn't believe he would begrudge my having it. We had both come up with that name, but it didn't matter to Slash. It felt like he snapped up "Road Crew" just so I couldn't use it. My pal Slash.

But it didn't matter because word of my new band got out through a kick-ass press kit that was making the rounds. We had write-ups in all the major metal magazines: *Kerrang!, Raw, Circus,* and *Hit Parader.* Davy and I were even booked on MTV's popular *Headbangers Ball.* Everything was going as planned.

A couple of shows were lined up; the first was in New York at the Limelight. After the gig, Nicolas Cage introduced himself to me. He's one of my favorite actors. He gave me his room number at the hotel he was staying in and told me to swing by. When I arrived later with a couple of friends, the people at the front desk would not let us up. Oh well, his loss.

Next up was a New Year's Eve gig at the Stone in San Fran. The show was fantastic, but I got so fucked up that night, I nearly died of alcohol poisoning. I'm sure the other guys in the band were thinking, "Who is this guy with the death wish? Is he too stupid to know he's skirting the edge, or is he too depressed to care?" The boys were getting really tired of my shit. But after only three months as a band, we had four major labels interested in signing us. A showcase was set up for one of them. Two men and a young woman had been sent to represent one of the record companies. We invited them to the studio and even had the event catered with great food and fine champagne. Everyone was super-nice and professional. We played a blazing set for them, and it was clear that they were impressed.

We were sitting around and having a few drinks when one of the execs posed an awkward question: "Steven, I must admit we have one concern. What about the drugs? I've read you've had a serious problem. How are you with that now?"

Without blinking an eye, I reassured her: "That's behind me. I'm clean." The three of them nodded. The entire evening went incredibly well, and they told us they were in agreement. They were going to add us to their roster of bands. We promised that they could count on us to work our asses off and do whatever was necessary to succeed. I invited them back to my house, where we could relax more comfortably and sign off on the details of our deal.

We drove up to my house and had barely made it to the gate when we were distracted by the loud clanking of an old car. The driver pulled right up in front of us. She was a strung-out, wiry young chick. She got out and handed me a cigarette pack, God-fucking right there in front of everyone.

I was dumbfounded. She saw the fury in my eyes and immediately took off. But the damage was done. It didn't take a rocket scientist to figure out what had just gone down. The record execs pretended to have forgotten about another pressing engagement they needed to attend and politely excused themselves.

I just stood there looking up at the sky. "Why? Fucking why me all the time? Why?" I turned to my band. "I am so sorry, you guys."

Ashley shook his head and said, "Fuck this." I went into my house alone, cursing myself. The guys packed up their shit and were gone the next day.

I so feared being alone that I asked our roadie, Rocko, to stay around and act as my personal assistant. I offered him a salary of a grand a week, which he happily accepted. We never really hung out or anything, but he would pick up groceries or dope, or give me a ride whenever I needed. I set him up in a spacious extra bedroom in the attic of my home.

A month later, another girl had moved in with me. She was a hot model named Analise. One evening, I was high as a kite, relaxed as could be. I had the fireplace going, and I was watching cartoons on my favorite network, Nickelodeon. Analise was in the shower. All of a sudden I heard a loud thump, and then a scream. "Now what?" I thought.

"Steven!" Analise yelled. I jumped up and ran into the bathroom to see her pointing up at the ceiling. "Someone's up there. I saw a gleam, a reflection through the crack of the molding." Later I found out that Rocko had drilled a hole through the ceiling and was videotaping Analise and other unsuspecting houseguests.

At the time I thought, "What the fuck?" Sure enough, there was a peephole in the ceiling of the bathroom. Just then, Rocko comes walking in, nonchalant as ever. I put two and two together and got right in his face. "You motherfucker. Get your shit and get out, or I'm calling the cops." He didn't even respond; he was out of there in ten minutes.

## ≁ DON'T FUCK WITH THE BOSS ≁

I continued to hole up in my room, completely adrift, doing my thing. I don't know how (maybe through Cheryl), but my mom discovered that the band had cut me off completely and that I wasn't going to be receiving any more checks from Guns N' Roses.

Mom really went to bat for me. She contacted a top entertainment lawyer and proceeded to sue the band. I had given her the go-ahead and was all for it, but I was barely involved. When I had to take the stand, my nerves were shot. I kept a small stash in my pants and every chance I could, I went to the bathroom for a few hits. As a result, I delayed proceedings more than once. But the jury liked me. They believed I was honest and candid, because I was.

During the course of the proceedings, they dragged the whole band in. Can you imagine how I felt watching Axl and Slash speaking out against me? Axl and Slash took the stand and came off completely cocky and arrogant. You stupid fucks. Thanks, boys. Your condescending attitude was one sweet gift to me. You thoroughly alienated the jury and I was awarded $2.5 million in damages and regained my 15 percent in continuous royalties.

Thanks, Deanna. Don't fuck with a lioness guarding her cubs.

In 1994, with financial security locked down for the first time in my life, I freely entered rehab in Arizona. There, I was roommates with Layne Staley of Alice in Chains. All we talked about was partying. Layne talked about his woman a lot and showed me some pictures that she had taken of him. They featured a naked Layne in a shower stall with a needle in his arm. All the pics were taken by candlelight. All this drug and party talk, as well as the pictures, drove us so crazy we couldn't resist the urge to party. So we took the fuck off after only a couple of days, leaving the facility behind. He went his way to score some dope and I went mine. So much for that round of rehab.

I returned home, and eventually Analise and I split up. Again I was alone in my own personal misery. Any letup in partying and I'd feel like my world was crashing down all around me, so I continually medicated myself. I was alienated from the world, and the only person I saw regularly then was my finance guy, Josh Lieber, who later turned out to be a complete scumbag. The bastard had me trusting him for years. My mom and dad trusted him too, but he screwed us royally.

## ⇥ SCREWED BY THOSE CLOSEST TO YOU ⇥

Originally, Lieber impressed us by discovering that the band's accountant had stolen $80,000 from me. The woman who handled the band's finances went from living in a small apartment with three kids to suddenly having a luxury car and a big extravagant house. He wrote a letter to her, which pretty much read, "If you don't have this money on my desk by morning, you are going to jail."

I felt secure with Josh and made him my adviser, even though he was charging me $20,000 a month. All he really had to do was pay my bills, and I paid him more per month than the total on all my bills combined. I didn't even want to know about or be involved with finances, and he knew it. What a fucking idiot I was.

I just wanted money when I needed it. He could see that I was always fucked up and clueless, so he must have thought it was a sweet situation. Who was going to find out he was bleeding me, getting rich off me? He figured I'd die soon enough, so he set out to gouge me with a vengeance while he could.

"Stevie, I need you to sign this," he'd say. Then I'd ask, "Oh hey, can you give me a thousand dollars?" He set me up with an ATM card that allowed me $300 from the machine daily. You'd think $300 would be good enough for one day, but almost every day I found myself walking into a bank branch to withdraw more than the ATM allowed. I was buying everything that I could get my hands on, coke, heroin, pills, weed, whatever.

The final time Josh cheated me, he got me good. He had me sign a check and claimed that it was for taxes. He said, "If they don't ask for it after a while, you can just keep the money."

I was so naive. "Oh, okay, great."

A few months later, I asked him, "Hey, what ever happened to that tax money?" He said there never was a tax money issue; I never signed anything of the sort. "Steven, you know how you are. You were probably high and signed an autograph or something." I was immediately pissed. What!?

How dare he try to dismiss the matter by saying I was delusional? My driver at the time was a woman named Mary. Thank God she was present when we had the meeting in question. She knew from the beginning that I had signed a personal check and felt it was cagey. She reminded me that she had originally brought it up to me because she thought it was suspect, but I wouldn't believe that Josh was being dishonest. He lived like a king off me for years, but after I had everything audited by an independent service, he was out. No court case, no fanfare, no threats, just count your blessings that I let you drain me for this long and get the fuck out of my life.

## ⚞ CHEATING DEATH ⚟

With Lieber gone, the coffers were suddenly wide open. I upped my intake of everything and ended up barely cheating death again. I drove my Jeep to get some heroin, and as soon as I scored I pulled alongside the road and attempted to shoot up. I couldn't get it in a vein, so I just shot it randomly in my arm. I resumed driving to get some sodas and beers. I turned the corner, and the next thing I know, I'm crawling up the sidewalk in a dusty haze. Some guy ran over to me. I looked up at him and he was just going on and on. "Oh my God. It's a miracle!"

I had sideswiped four parked cars before crashing between two others. The front end of my Jeep was pushed in past the windshield. I escaped with a bump on my forehead and a little cut on my eyebrow. While the eyewitness was going on about my good fortune, I passed out on the sidewalk and woke up in the hospital.

Later I was charged with driving under the influence and lost my license. But when you're an addict and you have places to go, having or not having a license doesn't really slow you down.

## ⚞ CLEANING UP MY ACT ⚟

A court order gave me three months to turn in a clean urine test. I knew I was very fortunate because they could have stuck me in jail. I felt that even I could pass a drug test given that much time. I figured getting out of L.A. would give me a better chance to clean up, so I rented a house in San Francisco, where I intended to detox (at some point). The first month I partied insanely. My good friend Steven Sprite accompanied me. Steve didn't party. He didn't want me to spend all my time in the apartment and occasionally talked me into going out. I'm so glad he did, because we went to a local bar where the center of attention was a blond-haired bombshell. This was a girl I absolutely had to meet. With my usual confidence I approached her. Her name was Cherry. We went back to her place and hung out a bit. We exchanged numbers, and she eventually became my girl of choice, a special friend for life.

While in Frisco, I also got to hang out with my ex-bandmate and local resident Davy Vain. Davy has a big heart and had forgiven me for fucking up our band's recording contract. One night we were on a rooftop in the city and found ourselves sharing the evening with Linda Perry, former 4 Non Blondes front woman and current A-list producer. I thought she was so hot.

Being in Frisco was a blessing in disguise. I had rented a room in my Calabasas home to a friend of my brother, Weasel. Just before I left we had acquired some really good pot seeds. I gave him a bit of cash, and he fixed one of the closets up as a green room. He was familiar with the complete growing process. The walls were layered with foil and he used special equipment, such as timers and growing lights. It was a convenient situation for all. Weasel could live in a great home for low rent, and I'd be able to smoke some choice weed when I returned. I couldn't imagine anything going wrong with this arrangement.

Then I received an urgent call from Weasel. Cops had come pounding on the door. They had obtained a warrant and searched my house. When I asked why, I was horrified by the explanation. My ex-assistant, Rocko the pervert, introduced to me by Davy, had been living in Santa Rosa. Now this terminal whack job, who had tried to videotape unsuspecting people in my bathroom through a hole in the ceiling, had just been arrested for suspicion of murder. What the fuck?

Apparently, he was kidnapping girls, drugging them, and then videotaping the unconscious victims in various positions and sex acts. The drugs he was giving them were so powerful that the girls would awaken on the side of a road somewhere with only a nagging suspicion that something had gone wrong. I was blown away to learn that the cops had found a badly decomposed body in his yard!

Rocko must have eventually fucked up on the girls' drug dosages, because all of a sudden one girl did remember. Then another. They gave the police enough information to lead them to Rocko.

They raided his home and found dozens of videotapes containing the graphic rape footage. They also found pictures. Among the motley assortment were pictures of my ex-girl Analise and my mom, Deanna. One had been taped in the shower and the other on the toilet!

Rocko had been spying on us and taking pictures the whole time he had been staying with me. I felt so terrible that I had allowed such a monster under my roof. Especially for my mom and Analise. Their privacy had been horribly violated. But I thank God that his madness fell short of doing anything worse to us.

Here's the biggest kick in the balls: Rocko told the cops that he was able to finance his disgusting habits because he lived with and worked for me. I told them that I never had any sort of social relationship with the guy. He was a gofer for me and that was it.

During the routine search of my home, the cops discovered our growing room. We had three or four plants in midbloom. In court, Weasel and I both pleaded the fifth. He was just renting a room there and had no idea what was growing in the closet. I was away for two months cleaning out, unaware that such mischief was happening in my home. Our strategy worked and we got out of it without any hassles.

# 19

# ROCK BOTTOM. AGAIN.

From time to time I would drive up to San Francisco to see Cherry. She wasn't my girlfriend, and I knew she saw other guys. One time I got sidetracked on the way with Laurie, a woman who was cutting my hair at the time. We were hanging at her place and I was just smoking a little rock. Now, Laurie smoked speed and she was entirely spun on the shit. She had her stereo cranked and kept on playing it into the night. To make matters worse, the music was that techno shit. I had a bottle of Valium and I had already taken eight V's to combat the noise. Even though I moved to another room, I could hear it through the walls, a constant annoying *thump, thump, thump.* There was no way I would be able to sleep so I had to get out of there. At five A.M. I took off. I hopped in the Bronco and headed for Cherry's. Not a good idea.

225

I was fucked up from the pills, swerving on the road. Cops pulled me over somewhere around Bakersfield. I didn't want them to confiscate my last ten hits of Valium, so I quickly popped them in my mouth. Next thing I remember someone's shaking me, yelling, "Adler. Adler!" I was lying on the jail cell floor. I had been there all day. They let me out, but it was pitch-black, and I was in the middle of nowhere.

Two other guys who had been released at the same time gave me a ride to the bus station. I was freezing, shivering uncontrollably. I went into a gift shop and stole a shirt to help me warm up. The security guard saw me and tackled me. He was this huge fat guy who proudly said to me, "I may be big, but I'm fast."

I told him, "You idiot. You may be fat but you're an asshole. You just caught someone who's on eighteen Valiums."

He called the cops, and it was the same guys who'd picked me up earlier. They drove me down the road two blocks and let me go. This time they just told me, "Get the fuck out of our town. Now." I got my truck out of impound and took off. If I had just gone to Cherry's none of this would have happened.

## ⊰ HARD COPY ⊱

By 1996 I was hanging with Cherry more and more often. I just enjoyed her company without having to be high. I stopped partying almost completely, had gained weight, and looked healthier. The TV news magazine *Hard Copy* caught wind of my recovery from addiction, and they contacted me, wanting to do a story. I proudly accepted. My brother Jamie was quickly establishing himself in Hollywood as a talent agent. He contacted a nearby club and arranged for me to play with a band that would be performing there. This allowed *Hard Copy* to get some shots of me playing live. While I was there, I was struck by the aura of a man sitting at the bar. He was wearing a huge sheepskin jacket, and as far as I was concerned, he had "it." He introduced himself as Steffan Adikka. He was a musician who was in a band with the exceptionally talented Gilby Clarke. Gilby had replaced Izzy in Guns N' Roses in 1991 and subsequently left the group in 1994. My brother was booking bands exclusively for Billboard Live, which occupied the space of the now defunct Gazzarri's. So I asked Steffan and Gilby if they would like to form a new band. They lit up. "Fuck yeah!"

Steffan worked in a lingerie store in Hollywood. I was struck by an underwear line that spoofed Fruit of the Loom. The label said "Freaks in the Room." I thought that was so great. I asked the guys what they thought, and shortly after, that's what we christened ourselves. We became Billboard Live's house band, Freaks in the Room, playing every Monday.

## STERN'S FREAKS IN THE ROOM

To promote the band, I made an appearance on Howard Stern's radio program in New York. I didn't want Howard to tear me apart, which can happen, so I brought two porn stars with me in an effort to distract him from zeroing in on me too much. I rented a limo, and I brought Steffan and my brother along too. At least a half dozen times, I would have the limo stop. I'd kick the guys out so I could have sex with the girls. Then I'd open the door and invite the guys back in.

I love Howard Stern, and I'm pretty sure he knows it. That day Howard showed me nothing but respect. He seemed to think that I was the coolest guy. "How the hell do you get kicked out of Guns N' Roses for doing drugs?" he asked. He was impressed by my tattoos too. "This guy is hard. He's got tattoos on his hands." During the broadcast, I extended an open invitation to my former Guns N' Roses bandmates to join me at Billboard Live anytime. Imagine my surprise when Slash actually took me up on it. He joined us onstage for "Knockin' on Heaven's Door." But it wasn't the reunion I would have wished for. It was very awkward and strange. In fact neither of us could bring ourselves to say more than a few words to each other. Very uncomfortable, but in a strange way, somewhat healing. I love Slash. I hate Slash. I love Slash. You get the idea.

Freaks in the Room had been together for only about two months when it became obvious that it was not to last. One evening during a performance, the sound was incredibly screwed up. I could not hear anything. I tried my best to maintain a groove but all I heard was mush. Gilby stopped playing and yelled at me right there, on the fucking stage. "Get your shit together, you fuckup." I wasn't high, I wasn't drunk, the monitors were just a garbled mess and it was impossible to keep the beat. I yelled back, "Fuck you!" That was the last time the freaks entered the room.

I got myself an apartment in Studio City right off Ventura and Laurel Canyon, near Jerry's Deli, where the golf range is. A neighbor discovered that I was living there and stopped by one afternoon. He brought some beer and weed. He was a fan and also a guitar player. He introduced me to his roommate and sister, Debbie. The three of us became fast friends. Debbie would cook all the time and every day she would invite me over for something to eat. I could never turn down a home-cooked meal. I wasn't terribly attracted to her, but she was cute enough, your typical sassy Italian girl from New York. It was purely out of convenience that I started having sex with her.

Debbie had an odd relationship with her brother. He wanted to have a sex change. He already had had plastic surgery on his face and a boob job. I saw pictures of him before the operations. He had a real big nose and long curly hair, looking nothing at all like his sister. But now, after the procedures, he looked just like her. It appeared to me as though he was changing himself in her image. I thought it was a scary infatuation, kind of like Michael Jackson's obsession with Diana Ross. Regardless, he/she was a nice person and the siblings were close.

## ⇥ NEW DRUM SET ⇤

A few days before my twenty-seventh birthday, I went to New York with them. My sole reason for going was to ice-skate at Rockefeller Center. I also got wind of a 1968 North drum set that was for sale. This was a particularly rare collectible, with four toms up front, a floor tom, and a double bass. It was a classic without a scratch. What drew me to it was that fateful moment in 1980 when Slash and I pulled back the curtain on Nikki Sixx and London at the Starwood. Their drummer played an identical North set, only in white. Now, I was gonna have one of my own for only $2,000. I got the guy's phone number and assured him that I would be in touch in a few days to make arrangements.

We returned to L.A. and went out to the Rainbow to celebrate my birthday. Jason Bonham, son of John Bonham, the legendary drummer for Led Zeppelin, was there with us. We had a great time. We returned to the apartment, where Debbie invited me in. She had large boxes spread all about the floor. She exploded as the realization hit me. "Happy birthday!" she yelled and gave me a hug. She had bought me the North drum set! I was so thrilled and my heart swelled; I thought she was so generous. We had sex, then slept away most of the next day.

The next evening Debbie and I were sitting in my apartment watching TV. Suddenly, she stood up and knocked everything off the coffee table. "Huh?" Then she walked over and started knocking my awards off the wall. She lifted the stereo and crashed it to the ground.

She was about to drop the TV when I stood up, grabbed it, and yelled, "What the fuck's wrong with you?" It was ten thirty P.M., and the insanity raged off and on for another eight hours. She just kept acting like a complete psycho, incessantly screaming at me and calling me a loser. I tried to be calm, but really, I thought I had a heart attack that night caused by all the anxiety. All I could think was that she was doing this to set me up somehow. She'd never acted out of line before. It was freaking me out. My chest hurt and I couldn't breathe; I just lay on the floor begging her to stop.

Finally, I had to do something. Around six thirty in the morning, I grabbed her and forced her out the door. "You're going down," she yelled. Two minutes later she calmly knocked on my door and announced that she had called the cops, claiming that I had beat her up. Immediately, I grabbed my shoes and my hippie bag, got in my truck, and hightailed it out of there. I never went back. I couldn't understand what she felt she stood to gain from all this. I guess she wanted more from me than I was prepared to give, and sensing that, she reacted by lashing out at me.

I relocated to another condo in Studio City. A female friend named Lindsay was with me and helped me set up the new place. After we moved in, I was awakened by a knock at the door. I looked at the clock; it was six A.M. "Who the hell could that be?" I shuffled up to the door. "Who is it?"

"Adam" was the response. I assumed it was the maintenance guy from the building. I opened the door to find two police officers. "Are you Steven Adler?" one asked. Before I could answer the other officer said, "Yeah, that's him."

I could only say, "Hey, what's going on? What's up?"

The officer who knew me said, "Mr. Adler, you are under arrest for IPV, intimate partner violence. Your girlfriend said you beat her up."

I looked back at Lindsay and yelled, "What the fuck's wrong with you?"

"Steven, I didn't do a thing!" Lindsay said, and it was true. It was Debbie carrying out her threats to do me in.

## ⊰ COURTROOM BETRAYAL ⊱

I was put in cuffs and brought to a jail in Santa Monica. Three hours later I was released on bail. I walked into a lawyer's office across the street from the condo, not because I was referred to them, but because it was close. Big mistake. Always get a referral for legal counsel from someone you trust, and then check around, or ask the lawyer if you can chat with one of his satisfied clients. As I later learned, this attorney I hired was a complete asshole and thoroughly hated in the legal community.

That's all I needed. It was only a month before I was to face the judge. I had a short trial that involved so much insane bullshit from Debbie that I just tuned it out. I seriously just blocked out the trial because it was all fabricated crap to begin with, and I didn't tune back in until it was time to receive my sentence. I don't believe the jury thought I was guilty, and it felt like I was gonna get off. But I overheard my lawyer mumble something to the judge. I could swear he fucking said, "I'm sure he's done *something* wrong." What the fuck? I was found guilty—I'm still not sure of what—and sentenced. But it was a relatively light sentence, three months and a fine.

First I was put in a police station in a small town in L.A. County called Laverne. Christian Slater had been sent there for some legal trouble just a few months before. It wasn't that bad at all; I had a cell phone and cable TV and was assigned to washing police squad cars. I shared the cell with three other people and was locked up from eleven P.M. to six A.M. I'd be sleeping during that time anyway, so it really was a free ride. But back at the condo Lindsay was having a terrible time.

My brother Jamie was making life hell for her. Over the past few years, Jamie had become increasingly difficult. Maybe he felt he had my best interests in mind, but sometimes he would cross the line. I had no problem with Lindsay staying at the condo while I was incarcerated, but Jamie sure did. He would swing by every day and harass her. She was driving my Bronco around and Jamie called the cops reporting to them that she had stolen it. They pulled her over and handcuffed her, then discovered that she was living with me and that everything was okay. But Jamie was relentless in his attacks, forcing Lindsay to move in with a neighbor a few floors up.

On the weekends I was allowed visitors. You could go to a nearby park or the local library. Lindsay would bring me grub from Taco Bell or some other fast food I would be craving. Then I would bone her in the truck or we'd do it in some public bathroom. While I was back in the cell, I had a great idea. I called Lindsay and had her make a dental appointment for me (which was permitted) so I could get out for a while.

She picked me up at the jail and we went back to the condo. I scored some heroin and smoked it all afternoon. Soon it became night. I returned to the jail, but hours too late. It never takes anyone that long to go to the dentist. I was such a jackass.

Understandably, they wanted me to piss for them. But I was so high I couldn't even pee. I couldn't get a drop in the cup, and that was evidence enough for them that I was fucked up. They locked me up and the next morning a guard came and got me. I was put in the cell at the courthouse for eight hours. Then they placed me on the bus with all of the other inmates du jour and took us to the L.A. county jail. It was twenty-four hours later before I could lie down and sleep. Boy, I fucked up. This was the real deal, not the cushy arrangement I had before.

For some reason they asked me if I had ever seen a psychiatrist. I figured if I said yes they'd leave me alone, but if I said no I was in for hours of shrinkage. If anyone was ever preshrunk, it was me, so I said, "Oh yeah, all the time, ever since I was a kid." It was kind of the truth. Whether it was nosy school guidance counselors or the resident quacks they had at rehab facilities, I was always getting analyzed whether I wanted to be or not.

Well, I ended up being placed with all the crazy people who had the luxury of receiving medication three times a day. All you did was sit around on the cement floor. If you were lucky you could sit on a bench. On each floor were three pods, each holding about thirty people. The walls were transparent Plexiglas. It was my first time in a real jail.

I would get letters from Debbie. She pretended as though nothing ever happened, like she had never fucked me over. My God, she was a complete and utter psycho.

Earlier, when I spotted her going into court, I pleaded with her, "Tell them I didn't do what you said. What is wrong with you? You are ruining my life." The only thing that kept me going was dreaming about how much I was going to party when I got out. I had to serve the entire ninety days. It seemed like forever, but it was nothing compared to the time some of the others had to do.

I was put to work in the kitchen. I served breakfast, lunch, and dinner. It was the absolute best thing, because oatmeal and chocolate-chip cookies equaled money. I had these baggy pants with big pockets that I would fill with the tasty delights. When medication time came, I would trade the cookies for the other patients' meds. I got nearly everyone's pills. In the ward they served nothing but downers. They didn't want anybody getting all wired or hyper. So I was able to be sedated most of the time. Nothing was too powerful, but everything helped.

## ⊰ RELEASED ⊱

At last the day came when I was released back into the wild. I hadn't been partying and I'd gained twenty pounds. Jamie picked me up with one of his girlfriends. We went straight to the Rainbow. I got pasta and did endless shots of Jägermeister. I got so sick, I puked my guts out. The Big Spit; it was great!

Now that I was back home I took every opportunity to go insane and party. Lindsay and I would score dope up to six times a day. Sometimes the coke would make me a little too anxious, so I took pills like Valium to keep the edge off. Smoking coke and heroin, and then shooting the two combined, became a favorite buzz. My relationship with Lindsay was based solely on partying. There was no real connection other than our insatiable appetite for substance abuse. We really went over the top, tweaking our brains out constantly. Within a week, I don't think either one of us was in our right mind. I was in free fall, utterly insane. My little preemptive warning system had finally failed to go off, or I was too far gone to notice it.

## ⊰ FACE FRAPPÉ ⊱

When you're as tweaked out as I was, it's never a good idea to prep your own fix. One night, I messed up the dosage, then messed up my face. I was speedballing, injecting myself with heroin and coke. But in my shattered state, I must have injected myself with a horribly excessive amount. As soon as it hit my bloodstream, I collapsed on the floor of my bathroom. My body started going into convulsions, and what was worse was my head began slamming uncontrollably and repeatedly against the white tile floor.

It was the most horrifying experience of my life. Try as I might, even with all my strength, I could not stop my face from smashing down on the floor again and again. My lip split, my teeth cracked, and blood began to flow from everywhere. Still, I couldn't stop. On the floor, a thick bath towel lay inches away from my head. If I had just been able to get it between me and the tiles, I may have been able to lessen the damage, but my body was so out of control, I couldn't will myself to grab it.

Mercifully, the convulsions eventually ceased. I don't know how long I lay there with teeth shards and bits of my face littering the bloodied floor. The next thing I recalled were the sounds of soft voices in quiet conversation, a brief moment of consciousness when I realized I was in a hospital room, and then more blackness. This seemed to go on for an eternity.

Finally, after several days, I was strong enough to sit up in bed and discuss my options with my mom and a plastic surgeon. There would be several reconstructive procedures, working in tandem with an oral surgeon. I was fortunate; the damage was reversible. The only thing that wasn't reversible was me, because this was all I could think about throughout the conversation: "Now, if I could just score some dope."

I think my mom saw the hunger in my eyes that day, just the way Steven Tyler had me pegged years before. As I pretended to pay attention to the surgeon's strategy for putting me back together, she knew the truth was no one could ever put me back together. She knew I was hopeless. I was way beyond broken, and it was permanent. My need to escape that sad fact through drugs was so powerful that the only way out of addiction would be death itself.

## ⊰ SUICIDAL DESPERATION ⊱

When I got back home, my fellow train-wreck-in-waiting, Lindsay, was still prowling around the house. She hadn't weathered my absence well and was even more strung out than before. When she wasn't high she cried a lot and seemed extremely fragile emotionally. It didn't take long before she bugged the shit out of me.

One evening I just couldn't take it any longer and made the mistake of suggesting we part ways for a while. "Maybe you should go to your mom's house," I told her. She didn't respond, only quietly retreated to the bathroom. I heard water running in the bathtub, then I must have dozed off.

When I woke up the next morning, I called out for her but there was no response. I looked around the condo, but she wasn't there. I called the front desk and the building security to see if they had seen her leave, but they hadn't.

I was consumed with an odd, uneasy feeling. Where the fuck did she go? I walked into my bedroom and froze in the doorway. I hadn't noticed this at first, but there in front of me I saw sheets tied to the bedpost. The sheets were draped out the fifth-story window, fluttering in the wind. This was all I fucking needed. I walked over to the window, filled with fear. "Oh God, no."

I looked out, and there she was, her body motionless, sprawled out hideously in the bushes below. She must have tried to hang herself, but the sheets didn't hold. She was naked and I could see that her body was covered with multiple cuts, some of which I later learned were previously self-inflicted. Her head hung at an unnatural angle, her left arm twisted behind it. I ran to the phone and dialed 911.

As fate would have it, my mom just happened to be coming by that morning to check up on me. When she arrived, she saw teams of cops racing into the building. She told me that she never thought twice about whose condo the cops were rushing toward. She stayed with me through all the questioning. Fortunately, the cops found only a little pot and my bong in the apartment, which they basically ignored.

Lindsay survived, miraculously suffering only a broken collarbone. We got her into the hospital and made sure she had the best treatment possible. After she got out, I never heard from her again. I never really thought about whether this was good or bad. If I wasn't that into a chick and she decided to move on, then that was good for both of us. Besides, when you're doing coke and heroin, you really don't have the ability to give a fuck about anyone else. Loved ones can be sick, injured, in the hospital, in jail, and you don't go to see them, you don't even give enough of a fuck to call them. It's not that you're selfish; it's just that the thought never occurs to you. The drugs demand all your attention every waking moment, and then you nod out and wet yourself.

## ⇥ ONE OF MANY ODS ⇤

I myself escaped death while living in a house in Calabasas. I would meet my dealer at the Thrifty parking lot in Laurel Canyon. This one afternoon, I brought my dogs with me to get them out of the house. After scoring, I'd drive to a little side street down the block. Then I'd pull over and shoot up. This was a regular routine for me that had resulted in at least one disaster already. All I remember from that day is sitting there cooking the shit up in a spoon. I vaguely recall seeing two little kids playing ball across the street. After that, nothing.

Apparently I went into convulsions again. I was violently flopping around, rocking the truck, my forehead slamming into the horn while the dogs went crazy. The kids heard the commotion and got help. Paramedics arrived and had to break the window to get me out. Again, I awoke in the hospital, covered in charcoal. When you overdose, they pump your body full of charcoal. It comes out of your nose, your mouth, and your ass. You spit charcoal, you puke it.

Shortly after that, there was another incident where I nearly didn't come out of it. I had recently been prescribed lithium. What they didn't know was that I was allergic to the drug. The shit was fucking me up more than anything. It turned me into a zombie.

One night I had a strong craving for a Slurpee. I had plenty of money, but I couldn't find any. I tried to get out the front door, but I couldn't even turn the handle. Sluggishly, I picked up my drum stool and threw it through the window to get out. But when I hopped over the broken glass, I sliced my foot badly in the process.

Wearing only boxers, I made my way down the hill. I was craving sugar. I must have looked so scary, limping along the street in my underwear, bleeding. I entered a 7-Eleven store, where the cashier eyed me warily. I didn't have any cash so I stole a Big Stick ice cream; it must have been obvious as hell.

So I'm walking back, boxers hanging off my ass, bleeding like a stuck pig, sucking on my Big Stick, when a cop car pulls up alongside me. I look at them with my zombie eyes and say, "I don't feel so good. I think I'm sick. Can you take me back to my house?"

They eyed me up and down and said, "You're on your own."

They took off. The one fucking time I want to get in a cop car and they wouldn't have me! There's a lesson for you: if you don't want the police to pick you up, beg for their help.

I made it back to the house, and it took me forever to open my sliding glass door. When I finally did, I stepped inside and tried to close the door but just said, "Fuck it," and collapsed on the couch.

Two days later I went through this exact same routine. Wearing only boxers, I went to 7-Eleven for a Slurpee, only to discover again that I had no money. My old house, the one I sold to former MTV VJ Martha Quinn, was close, just up a big hill. So I made my way there and buzzed at the front door. Martha's husband was the greatest guy. He could have called the cops, but instead he just listened. I was a royal mess, swaying back and forth, slurring my speech. That lithium!

I said, "Please, I can't find my money. Can I borrow $1.25 to get a Slurpee? Maybe you could take me down to the store?" So he drove me down to 7-Eleven, only to find that the goddamn Slurpee machine was broken. That kind of sums up my luck and my life. Even with help, the deck is always stacked against me.

# 20

# HOW LOW CAN I GO?

## ⤜ MINOR MIRACLE TIME ⤛

I have no idea why, but two days later one of my lawyers called my buddy Steve Sprite and asked him to check up on me. Over the years, Steve had proven himself over and over to be my absolute best friend, and he's always there when I need him. He's been by my side for years, and I don't know how I'd manage without him. Maybe it was the lithium, maybe something else, but for some reason, I hadn't been able to eat for days. I was just miserably sick from an ulcer or something. I'd get incredibly hungry and order up a ton of food from Jerry's Deli—soup, sandwiches, knishes, mashed potatoes and gravy. Then I'd take one bite and get a horrid feeling in my stomach, reminding me that nothing would stay down.

I called a friend of mine, a woman who tended bar at a barbecue restaurant nearby. She swung by from time to time and helped take care of me. She and Steve took one look at me and were freaked out by the terrible condition I was in. Steve dragged me to his truck, a big piece-of-shit white Chevy that only he could get to start. I was so fucked up, I believed I was riding in a brand-new truck: I remember telling him, "When did you get this new truck? It's like yours, only nice." I thought it was so styling, a beautiful brand-new pickup, so I must have been delusional.

Steve got me to the emergency room of Century City Hospital and put me in a wheelchair. I remember being pushed through the doors. I kept repeating, "I didn't do any heroin, I didn't do any heroin." I was completely out of my mind. I was put on a hospital bed, where I just stared up at the light. Steve was worried that I might be in a lot of pain and told the doctor. The doctor calmly walked over to me and yanked the hairs on my chest to prove I didn't need anesthesia. He knew what he was doing, because I didn't even flinch. I had a large infected lump on my arm, the result of my being completely abusive with a dirty needle. The doctor cut into it with a scalpel, and all this green ooze came shooting out. The nurses had to move the person in the next bed to another room because it smelled so awful. Then something let go inside of me and I just went out. Later, doctors would determine that I had slipped into a coma.

At some point during the time I was out, I remember floating in a dream that was incredibly vivid. I saw myself lying on top of a giant turquoise phoenix in the middle of a desert, and to either side of me were giant turquoise stones shaped like the phoenix. My point of view shifted and I was now in the sky, looking down upon myself. A vision of a woman floated up to me from beneath the clouds. She was nude except for a small garment covering her midsection. Her hair fell all the way to the ground. But it wasn't exactly hair, it was hair that morphed into feathers, and those feathers formed wings. She flew above me and reached out. I was lifted into her arms and I swear I never felt so safe, so secure. We were hovering high above the earth as she turned and smiled at me. We suddenly shot up into the sky, where the light became blinding.

I don't know how long I was out or how deeply I had slipped into a coma, I just know that I caught a break, the biggest break of my life. Hospital records show I was admitted on April 19, 1996, and sometime between then and April 22, when I walked out against their wishes, a miracle happened.

The blinding light caused my eyes to flutter open suddenly, but the fluorescent glare blinded me and I quickly shut them to stem the searing pain in the back of my head. I was groggy, but I could feel the needles and tubes leading in and out of my arms and thighs. I thought, "Fuck this." I angrily started yanking them all out, but the machines around me set off warnings, urgently buzzing and ringing. Something was very wrong; parts of my body weren't working properly and this intensified my rage. A nurse ran in and shouted at me to stop, but I just stared at her. I had no grasp of what I had just done, had no idea what was going on.

Later that evening, a doctor performed a series of tests on me and told my family, who had come to visit, I was in stable condition and recovering. He shook his head and quietly slid the clipboard back into the metal sleeve at the foot of my bed. I looked around the room at my family. Mom, Mel, and Jamie were there, and that was fine with me. My mother climbed into my bed to be next to me. She was crying, a steady quiet sobbing. I asked her why. She explained how the doctor feared that initially, before my tube-tearing fit, I wasn't responding to their tests, and they determined that I could be trapped in a coma for the rest of my life. The doctor told her that even if I were to come out of it, I would not be able to use my right side. She kept shaking her head, crying, going on about how it was a miracle, how amazing it was that I was responsive and, according to their latest tests, almost fully recovered. I didn't come out of it, however, completely unscathed. I lost some control of the muscles on the right side of my face, and to this day I talk with a slight slur.

The doctors concluded that I did, in fact, suffer a stroke. You'd think this nightmare would be the final moronic chapter in my life. And you'd be perfectly right to assume this would scare me into cleaning up my act. You'd guess I finally suffered something so traumatic, I'd find the resolve to take better care of myself.

Guess again.

Within two days I was walking around, much to the amazement of the physician. They actually presented me with a plaque that said ANYTHING IS POSSIBLE.

*I promise you anything*
*Get me out of this hell.*

—"COLD TURKEY," JOHN LENNON

The past decade had been a long, hard road. Unfortunately, the stroke did nothing to deter my drug use. Subsequent shooting pains across my chest left me weak and worried. And while I didn't know if it was a heart attack, I did know I'd be using again.

As we entered the new millennium, I remained detached from the world. I rarely went out. Whenever I did, I found it difficult to enjoy myself. I would always return home depressed, immediately hitting the pipe or calling my dealer. A long line of unsavory characters slowly invaded my inner circle and soon became its sole inhabitants. With the exception of Steve Sprite, the one friend who would never let me down, all of the people I saw were drug dealers or drug users. Any time the phone rang it would always be one of three people: a dealer, Steve Sprite, or my honey, Cherry.

The exception was one call I received from a buddy who dealt in shady underworld business. He asked me for a favor. It seemed the law had finally caught up with him, and he was going to be doing a little time. He needed a safe place for his girlfriend to stay while he was behind bars. Would I mind letting her live with me for a few months? When someone who's probably connected to the mob asks you for a favor, there's only one answer you can give.

Her name was Sheila, and she was the daughter of a designer clothing mogul. Now in her midthirties, she found herself cut off from her family, a tragic result of her own pathetic drug addiction. I had to be careful of the way I'd budget for this extra member in the house, because at this point, Mom was overseeing my finances and she had me on an allowance of $600 a week. I would get the money on Friday, and by Tuesday or sooner, I'd be broke. I also had the use of my credit cards but was not permitted to withdraw cash with them.

The drug abuse shot my paranoia to new heights. With her boyfriend in jail and Sheila staying at my place, I was convinced that my condo was bugged. I believed that I was being watched by the Feds. Any time anyone would begin to talk about drugs, I'd hush them up or make them talk in whispers.

I began calling my mom every day for more money. I spent many mornings cursing her out over the phone. "You fucking bitch. That's my money. Give it to me." Withholding cash from me was entirely justified, but there was no way I'd admit it at the time. When addiction consumes your soul, you lose sight of yourself completely. I had become an evil fucking demon.

Mom had limited my credit card so it could only be used to buy food. So Sheila and I worked out a scam where she would buy over a hundred dollars' worth of items, like cleaning supplies and dog food, with my card, only to return it all shortly after, untouched, for cash. This went on for months.

Sheila and I had a number of sources for drugs. One was a terribly shy, mentally challenged man named Bob. When the going got rough, Sheila would perform sexual favors for him, and he soon fell in love with her. Unfortunately, she hated every minute with him, so it was only during the most desperate times that she was forced to call him, meet him, and come back with a fat rock for us to smoke.

Some mornings I would wake to discover one of my gold or platinum records missing from the wall and an unconscious Sheila spread out on the couch high on heroin. She hocked more than a dozen awards of mine. I would become so angry and often found myself heading over to the pawnshop in an effort to retrieve my lost treasures.

On one such occasion, I ran into a face from the past. Ola, Slash's mom, was browsing in the shop. "Ola?"

I was delighted when she recognized me immediately. "Steven! How are you? You look great. You cut your hair, I like it."

I was too embarrassed to admit that my short cut was a result of accidentally setting my hair on fire a few weeks before. Instead I just happily laughed. "Well, thank you."

She mentioned that she would try to have Saul call me soon. While that would have been great, I wasn't going to count on it. I shifted back and forth on my feet, embarrassed that she might notice the GNR awards on display at the shop. Fortunately I don't think she did.

## ⇥ GETTING OUT ⇤

I did venture out once or twice during this time. Steffan, a longtime friend, invited me to check out his band, Dad's Porno Mag, at the Whisky. The show was sponsored as part of an annual convention called NAMM, the National Association of Music Merchants. During the show, I got up and contributed backing vocals to a favorite oldie by the seventies band Dr. Hook and the Medicine Show called "Cover of the Rolling Stone."

Afterward, I was outside talking with some people when this big guy with a bigger grin approached me. Despite his expression, he was somewhat intimidating, and I wasn't sure what to expect. "Steven, huge fan!" he said. I shook his hand and just said, "Thanks." We'll call him Chuck and his girlfriend Kimberly, his friend Larry and Larry's girl Sue. They were all in their twenties and turned out to be sweet people.

Kimberly was a petite, attractive blonde. She said, "I used to be in love with you."

I barked, "Used to?" They all laughed. Chuck said he would not have known it was me had I not been introduced during the show. That was probably because my hair was so much shorter and my skin had taken on a sickly gray tint.

I realized it had been quite a while since I had been recognized in public. One of the last times was not pleasant at all; in fact, it was downright heartbreaking. I was in the frozen foods section of a local grocery store, stoned out of my mind as usual. A shapely female had caught my attention, and as I walked closer to her, she turned toward me.

It was Cheryl! She had barely changed and was as beautiful as ever. Before I could utter a word, she looked me up and down, put her hand over her mouth, and began to sob openly. I tried to say something, but she quickly turned and ran away. For at least a full minute, I just stood there in shock.

Being around Chuck and his crew reminded me of the friendly, enthusiastic fans who surrounded me during my heyday, and that was nice. Chuck proclaimed, "I've got over a hundred GNR shows on video."

"No way. I need to see them." A few days later I had him over. He was thrilled. He asked if he could bring some things for me to sign. No problem.

Sheila and I made sure to score plenty of dope before Chuck came by. Chuck arrived and handed me a box of over fifteen videotapes, each with custom-made GNR covers. It was the coolest thing ever. First, I popped in a show from the Felt Forum, New York. While the concert was playing I would duck into the bathroom to smoke out while Sheila remained, and when I returned, she would drop out of sight. We took turns. It must have been obvious as hell that something was going on.

## ⇥ DEVOTED FAN ⇤

It didn't take long before Chuck enthusiastically bombarded me with questions. Some I wasn't entirely comfortable with answering. He asked me about the contract that I signed in 1990, which I assumed was the probationary agreement carrying the $2,000 fine if I slipped up. "What? I signed something that said I couldn't party?" I wasn't really up to explaining everything to him and dismissed the whole matter by branding it "bullshit."

"Well, do you still party?" he asked.

"Only weed. That's my last vice." I lied.

Chuck had brought me a CD he compiled of all of the early Guns N' Roses demos, including the original version of "Don't Cry." Man, talk about memory lane. I hadn't heard these recordings in years. I popped the CD on and got behind my electric drum set, which was set up in my bedroom. First was "Shadow of Your

Love," one of the band's fastest songs. I played along and didn't miss a beat.

Among the items that Chuck had brought over for me to sign was an eight-by-ten photo of me at the Ritz concert taped by MTV. I said to Chuck, "Hang on." I ran into my closet and grabbed the pair of leather pants I was wearing in the picture. I wanted to give him something to show my appreciation for his devotion, much like Nikki Sixx had done for me with his leather jacket. Chuck was delighted, telling me, "This is rock history, the Ritz show, right here." I felt a temporary lift in my mood, and I embraced it. I had nearly forgotten what it was like to be treated like a celebrity and not a fuckup.

## ⊰ FORMER ROCK STAR–BE PATIENT ⊱

A friend of mine had gotten me Slash's phone number. I started to call him regularly but was always bounced to his voice mail. I left him messages, but unfortunately he never returned them. I guess I felt that if I kept calling, I would eventually reconnect with my old friend.

I continued with my daily routine, partying with Sheila, watching TV, eating only from time to time. Chuck called quite a bit, leaving upbeat messages, and one evening I invited him over while I had a number of other friends around. One of them was John Weissmuller, grandson of Johnny Weissmuller, the Tarzan from the thirties and forties films. He came by with two of his friends, a guy and a sweetheart of a girl they had just met.

The girl offered to buy food for us, so we called up to Pink Dot, the convenience store delivery service. While she was placing our order she gave them my address and phone number, then let out a laugh. She told us that the person from the store said that there was a note next to my phone number on their business computer screen: "Former rock star—be patient." Now, that's perfect. They certainly know me.

Chuck had brought over a number of eighties L.A. metal magazines, including *Bam* and *Mean Street,* all featuring my old band on the covers. As I was flipping through an old *Bam* article, someone shouted, completely out of the blue, "I wanna smoke some coke!"

Chuck asked, "Why in the hell would you want to do that?" The guy looked at Chuck, confused, thinking, "Why the hell not?" A few minutes later, I decided to kick everyone out, and nobody protested.

Several weeks later, Chuck talked me into going with him to the Rainbow. Getting me to go anywhere was no easy task. But I loved the Rainbow, and Chuck was persistent. Later that evening at the Rainbow, I bumped into Carmine Appice, a legendary drummer who had played with Jeff Beck, Rod Stewart, and Ozzy Osbourne, and had been in some great bands, including Cactus and Vanilla Fudge. Carmine was jazzed to see me. We grabbed a table and had dinner, during which he reminded me of a show one time when he was asked to introduce GNR. He mentioned that he couldn't believe what Axl had done to Guns N' Roses, getting rid of virtually every original member over the years. We just shook our heads.

The best part of the evening was when he told us stories of being in Vanilla Fudge and completely surprised us by telling us it was he who wrote the huge hits "Hot Legs" and "Young Turks" for Rod Stewart. Then I told Carmine my Rod story. Shortly after he had married supermodel Rachel Hunter, word somehow got out about the brief dalliance I had with her. Rod called me, and I was completely stoked to be speaking with one of my rock 'n' roll heroes. "I'm going to kick your ass," he said threateningly.

"Cool. Fuck, you're Rod Stewart. I'd be honored to have you kick my ass." Carmine laughed heartily. It was a great evening, one of my best in a very long time.

ater that week, back at the condo, Sheila introduced me to a man I truly admired, Patrick McGinnis. He was a top lawyer but also one of the most honest people I had ever met. He had recently beat cancer, which had given him a new perspective on life. He gave me the lowdown on all the old lawyers I told him I had hired and was not surprised by the trouble many of them had caused me. I was reassured that I could count on him for anything. He was filthy rich, so I figured that I would never have to worry about him ripping me off.

One evening, he invited me out to Chasen's, this swanky restaurant in Beverly Hills. The rugs were plusher than plush and it reeked of old Hollywood. I opened the menus and my jaw dropped. A small salad was like $20. Patrick offered to pick up the tab, but I didn't want to take advantage of his hospitality, so I opted to just have a drink instead. There was a live band playing, and Pat must have sensed my taking them in, because he asked if I would be interested in joining them for a song or two. I smiled broadly just at the thought of it. Well, Pat had serious pull all over town, and I'm sure he knew the owners or management at Chasen's, because a moment later I heard: "Ladies and gentlemen, we have a treat for you: from Guns N' Roses, please welcome Steven Adler." The place went nuts. I got behind the kit and we rocked out to "Honky-Tonk Women" by the Rolling Stones.

Pat was overjoyed and bought the entire room a round. I felt on top again, but the emotion was fleeting. This is so sad, but the pangs of my addiction were getting the best of me, and I had to leave abruptly, much to Pat's dismay. "Steven, c'mon, do a few more songs," he begged.

"Sorry, Pat, got another commitment."

The drugs were beckoning again, and I had to get home. I knew Sheila would be there, but despite her company, loneliness was my true companion. Sheila was around out of necessity. She was a good girl who ended up traveling a road that can bring out the worst in anyone. Her very presence encouraged drug use, and I prayed for the day she'd leave. Although the possibilities of a new love in my life seemed remote, I still yearned.

### ⚔ CHERRY POP ⚔

When the loneliness became overwhelming, I would call Cherry. She still lived in San Francisco, but I foolishly kept her in my heart. I was so obsessed with the thought of her that I ordered six copies of the 1996 holiday issue of *Hustler* magazine that I was told she was featured in. Unfortunately, when I got the magazine, I realized it was the December '96 issue that showcased her, not the annual holiday special.

One afternoon I had the wild notion to just pop in on her. I grabbed a flight to San Francisco and surprised Cherry at home, but she was not at all pleased to see me. I excused myself to go to the bathroom, and when I came back, I stopped short in the hallway to overhear her and her mother discussing how I looked like a walking corpse. I kept hinting that I wanted to be alone with Cherry, but she stalled, and when her mom finally left, Cherry said she would not have sex with me. She finally agreed to give me a blow job but insisted that I wear a rubber. I had unprotected sex with this woman a thousand times. Did I really look that bad? I must have.

I was genuinely pissed when she suggested that it would be best if I leave. It was pouring rain outside. She offered to take me to some local commuter airport, and I just said fine. She dropped me off out front and as she drove away, I discovered there were no more flights that day.

I walked along the side of the highway, down the long-ass hill that led me to a windswept port, a lost soul with no money. I stopped in several motels along the way, and none of them had a single room available. I was drenched and ducked into a nearby Denny's, taking a seat in the lobby. A waitress approached me and asked, "Can I help you?"

"Yes." I smiled. "May I use your phone?"

"Sir, there is a pay phone right behind you." I explained that I didn't have any money. Her eyes narrowed and she turned away. A minute later the restaurant manager approached me.

"Sir, I need to ask you to leave."

"But it's raining outside."

"There is a homeless shelter a few miles down the road."

As I walked out of that Denny's in the pouring rain, I realized that I had truly, miserably hit bottom.

When I got home, my addictions just took over, and I continued to party. I thought it was amazing that I hadn't killed myself yet. When I smoked coke, my heart pounded viciously. One hit was never enough, nor twenty. I became so wired, I was paranoid *and* schizophrenic. I'd lock myself in my bedroom or bathroom, but this would only intensify the paranoia. I was so spun on the shit, a mere knock on the door made my pulse rate skyrocket. The answer for me was insane: I increased my usage.

The cycle was brutally nonstop, with incredibly abusive bouts yielding to pathetic attempts to clean up, which only served as an unintended way to set myself up for the next free fall. I didn't realize, never consciously grasped, that I was locked into this destructive routine. I just lived it, up and down, up and down, with occasional disruptions caused by my stroke, some cardiac incidents, and an uncontrolled fit that would leave me so mangled, I couldn't possibly reenter the cycle for a while. But as soon as I felt better, ususally one notch above foot-in-the-grave status, it was time to party. The resultant depression became so suffocating that I didn't even know it was a stage; I believed this was all there was to my existence. This intensified the drug abuse until I crashed and burned more disastrously than before. Those were the only variants: the severity and extent of the wreckage.

Beginning in 1990, when I was kicked out of GNR, this was the way my life degenerated for nearly two decades. I had literally created my own personal lost generation, a high that ended up making me feel so low, nothing could bring me back. Eventually, death seemed an attractive option, or more precisely, an alternative to the satanic treadmill I had created for myself.

# 21
# NEW BAND *and* NEW LOVE

### ⊰ SLASH'S SNAKEPIT ⊱

After parting with GNR, Slash had formed his own band, Slash's Snakepit. I had an opportunity to see them as Jamie, hitting his stride as a promoter/manager, could now get us into any club in L.A. He invited me to check out Slash at his show at the Whisky. Before the show, I met my old friend Taylor, who knew Slash too. She had been a part of the band's circle since the beginning. I entered the upstairs area of the Whisky, where a section was roped off for friends of the band.

Someone yelled, "Hey, Adler!" and I saw Slash's head turn. He saw me and immediately got up and approached me, looking surprised. "I just figured it was Jamie who was here, then I saw you," he said. We hugged, and I took a seat next to him. "So how ya been?" he asked.

As usual, I didn't think twice about sweetening the truth. "Great! Oh man, I'm playing everywhere. There's so much going on." I wanted him to know I was psyched, ready for action. I was thinking that it would be so great to join Slash onstage.

"So what songs you guys doing?" I asked. Slash suddenly became incoherent and mumbled something. I know him well and was disappointed that he had no interest in inviting me to join his band for a tune. I turned and looked for Taylor and saw that she was still behind the ropes. Security would not let her in, so I asked Slash, "Hey, let Taylor in?" He refused. I couldn't believe the nerve of this guy. I stuck a middle finger in his face and stormed off. I was so hurt.

During this time, Jamie started coming around more often. We were getting along so well we decided to move in together. We rented a house in the Hollywood Hills. He had the downstairs to himself. One of his buddies, former porn star Buck Adams, showed up from time to time and we became fast friends. Although he was now a producer in his industry, it didn't quite pay the bills. So Buck had a side gig painting houses.

## ⊰ FUTURE LOVE ⊱

One evening Adams, Chuck, and I hit the town for a night of drinking. On the way down, Adams and Chuck were going on about how they hoped to meet some ladies. I felt the same way, but I said, "Y'know, I don't even want to think about it, because when I do, it doesn't happen."

We arrived at the Rainbow, where a friend told me there was a cute girl from Argentina who was dying to meet me. He described her as a petite, well-endowed brunette wearing a tight Slash's Snakepit shirt. I was flattered but blew it off for the most part. We headed down to the Cat Club, located next to the Whisky. My old buddy Slim Jim Phantom had opened the place a few years before. We went out to the back porch, and there she was. She was a vision of loveliness, cute, innocent, and very sexy. At the sight of me, she became terribly excited, bouncing up and down with a radiant smile. Her name was Carolina, a twenty-two-year-old Argentinean. As attracted as I was to her, the lure of the drugs forced me to retreat home. Before I did, however, I gave her my number and told her she could swing by my house the next day.

Chuck spent the night, and the next morning I asked him if I could borrow his car, telling him I was going to 7-Eleven. Just before I left, Carolina phoned. I gave her directions and invited her over. I scored some shit, and when I returned, Carolina was waiting outside. I smiled and greeted her. I brought her in. Chuck was still asleep. I woke him and asked him to help keep her company; I was going to party in private for a few minutes. I told him, "Just keep her occupied."

"What do I tell her?" he asked.

"I dunno, tell her I'm cleaning." And that's what he did. Guess I was cleaning the bathroom for three hours.

That evening I invited Carolina to stay. In fact, after getting to know her over the next few days, I knew that I wanted her to stay with me permanently. Besides being a virgin, she was truthful and down-to-earth. There were no head games to play with her. I liked her honesty, her lack of ulterior motives. Keeping her around would be no easy task as she was in the States on a visa that was to expire the following week.

As the months passed, Carolina and I fell in love. Her feelings for me helped to disrupt my abusive ways, and I started to party less regularly. I found myself going out more, pulling myself out of the darkness with Carolina on my arm. Caro, as I came to call her, must have been heaven-sent, because by this point, the partying needed to stop, and she gave me the strength to at least try. Slowly, I began to feel whole again. We did loving things together, like buying a dog, another Irish pug. We named her Shadow after the old GNR song "Shadow of Your Love." She became our pride and joy, our little baby.

This could have been one of the most joyful times in my life, but for whatever reason, my family did not approve of Caro. Jamie in particular was a complete pain in the ass. He would be civil around us, but then he would turn around and tell my mom and all my friends terrible lies. He would say mean, destructive shit, like Caro was using me just to stay in the U.S.

Finally it got to the point where there was no way he and I could live under the same roof. Caro and I moved into my Studio City condo. But Jamie was persistent in his attempts to drive a wedge between me and my girl. He sank to an absolute new low when he had one of his Spanish-speaking friends call Caro's dad in Argentina and fabricate an insane story, saying I was pimping Caro out to turn tricks on the street and I would use the money she brought in to buy drugs.

Thankfully, Caro's father sensed this was a load of crap and immediately called his daughter. When Caro told me what her father said, I wondered just how far Jamie was capable of going. I knew in my gut that Jamie might be vindictive enough to call INS. I just couldn't risk his talking to some immigration officer, so something had to be done.

## ⇥ FOREVER GIRL ⇤

I called Chuck and told him that Caro and I got married. We in fact hadn't, but I insisted he post it on the web page. Chuck had started a small website for me, "The Official Steven Adler Fansite." It had grown quickly, averaging over a hundred hits a day, and I knew how news could explode on the web. I figured once my brother read this, or heard about it, he'd back off. The news spread quickly, and I even got a call from the *Hollywood Reporter* congratulating me, stating that they were going to print the wedding news. The plan must have worked because Jamie finally laid off.

And honestly, as far as my true feelings for Caro went, we may as well have been married. A few days later, my buddy Steffan was hanging out with us. I overheard Steffan ask Caro when she was planning to leave and go back home to Argentina. She never responded because I answered for her. I told Steffan, "Never." Caro was here to stay. I wanted her by my side forever.

Maybe my brother mellowed after realizing I truly loved this woman, because in August 2002, Jamie arranged for me to travel to Argentina to perform with Gilby Clarke. I saw this as a gesture on his part to gain some measure of redemption. The plan was for me to make an appearance during his band's encore and be paid handsomely for it. Unfortunately, Caro would not be able to travel with me, as the visa circumstances would have prevented her from returning to the States with me.

In Argentina, we arrived at the hotel, where throngs of people awaited us. We got out of the van, and our entourage had to cross the street surrounded by the masses. I was immediately overwhelmed, so joyful that I was being received this way. I took my time, signing everything the fans shoved in front of me.

Gilby had already crossed the road and was yelling at me, "C'mon, man. Let's go." He was very annoyed with me. But y'know what? The unconditional love I felt at that moment was helping to heal me. The trip was definitely therapeutic. It took me a full hour to reach the other side of the road. I rejoined Gilby, who was just irrationally pissy. Once inside our rooms, it was hard to ignore the chanting outside. "Steven! Steven! Steven!"

## ⇥ THE IN-LAWS ⇤

That evening, I got to spend time with Caro's family. I met her dad, who was a truly wonderful person and a well-respected businessman in the city. I just fell in love with Caro's lovely sisters. Her entire family was so hospitable, gracious, and loving. I could see why Caro was the way she was, so loving, so secure.

When Gilby's band played that evening, they received a very nice response. When I was brought on for an encore, the entire place exploded. The fans charged the stage in a total frenzy, and I again felt this wonderful rush of affection. It's been said that addicts merely replace one addiction with another during their life. Imagine if I could replace drugs and self-loathing with love and self-respect; that would be one hell of a swap!

We did another show with a similar fanatical response, plus Gilby and I did a TV interview. It was the absolute best experience. I had signed a young fella's arm our first day there, and when I saw him again, I was blown away. He had my signature tattooed on his arm. God bless you, man. God bless Argentina.

Sadly, the boys in Gilby's band, and Gilby himself, were upset with me. No one spoke to me the entire plane ride home. When we arrived at LAX, I hoped to catch a ride with one of the guys. They took off so fast, leaving me in the dust before I could even ask. I had to hail a goddamn cab. But I had been through this countless times before: amazing highs and lows in a single day.

It kills me to admit this, but as happy as I was, I could not escape the lure of the drugs entirely. I was back in my old condo, close to my old connections. Caro would get terribly upset with me, and that helped me make one of my boldest decisions yet. I was going to buy a house in Las Vegas, right down the street from my mom, who had moved to Vegas several years earlier and loved it. I felt this would be an effective way to distance myself from all the temptation.

## ⊰ VEGAS—PERMANENT VACATION ⊱

Once we arrived in Vegas, I gleefully hugged my dad Mel, my brother Kenny, my nephew Max, and of course my mom. I was proud of Kenny, although we seldom had much contact. Kenny and his wife had raised my sweet nephew Max, a loving, handsome boy, the pride of our family. We piled into the car and Mom took us to our new home, in the upscale Las Vegas Country Club community. We moved into a beautiful three-bedroom habitat. At my request, we had the interior redone in wood paneling and mirrored walls, and Mom made sure the fridge was stocked with my favorite snacks. I was upbeat about this transition; it was the right move, and it was going to work out.

As the months passed, I found myself battling the devil again, getting bored and restless. So I did what any self-destructive loser would do in this situation: I sought out a new drug connection. I had a bike, and every day I would pedal out to pick a little something up. I wasn't going near heroin again, but I had no problem hooking up with the other mistress, crack cocaine. It was my private little way to ward off the terrible monotony that permeated my life now.

Caro is the most patient person I know. She's been exceptionally good to me. We could argue an entire day, but when I awake the next morning, she's still there to say "I love you." So many times I felt that I wasn't doing much for her, so I was happy when she started to make friends in Vegas and was able to get out more often. Whenever we could, we would go see a concert. I always managed to score some backstage passes, something I know she enjoyed.

On New Year's Eve 2002, I was invited to check out former Mötley Crüe vocalist Vince Neil's band. They were great, and I was so happy when Vince himself invited me onstage to perform an impromptu version of Zep's "Whole Lotta Love." Vegas resident and Quiet Riot vocalist Kevin DuBrow joined us too. Kevin was so much fun that evening. He had the greatest bluesy voice and his trademark striped mike stand and ever-present suspenders. Kevin will be sorely missed; he left us way too soon, a victim of a cocaine overdose in 2007.

Afterward I was approached by an excited fan, Ryan. Now, Ryan was a nice genuine fellow, and he brought Caro and me a round of drinks. He told us he owned a very successful ticket service. We exchanged numbers and agreed to get together soon. We started hanging out often, with him taking Caro and me out regularly. As we got to know each other better, Ryan suggested a way to get me playing regularly again. He was close to Slash's Snakepit guitarist Keri Kelli. Ryan wanted to get us together, potentially to form a new band. He also offered to be my temporary manager. I happily agreed and a real bond began to develop.

I gave Ryan the names of some people I would have liked to work with, and he did his best in securing them. I had suggested another former Snakepit member with whom I was already friendly. This was Eric Dover, the vocalist. I also wanted my buddy Steffan to join us on bass. They turned us down, however, much to my regret.

So Ryan went to work and tapped the old L.A. band Love/ Hate's vocalist, Jizzy Pearl, who was currently singing for Ratt. This built some momentum, and soon, Brent Muscat, my old buddy from Faster Pussycat, was joining us on rhythm guitar. Keri's buddy Robbie Crane completed the lineup.

I will always owe a debt of gratitude to Ryan for helping me put my band together. I had the idea to tour, solely playing the songs that I had helped write for GNR so many years before. We would perform nearly all the tracks from the now legendary *Appetite for Destruction* album. The name for our new band: Adler's Appetite.

I flew to Orange County, California, to join the guys for our first rehearsal. Keri had a house, as well as his own studio, in the OC. The guys had already learned the songs. It was magic and felt so right, particularly because note for note, Keri sounded like Slash. Brent took the loose rhythmic approach that I loved so much about Izzy's playing. Robbie pounded the bass like the pro he is. Jizzy, well, Jizzy hit those high notes with the kind of vengence that made Axl famous.

Ryan booked our first show in Arizona, which served as a warm-up gig for our official debut. It was in Boulder Station Casino at the Railhead Lounge. Word got out to the press, and *Rolling Stone* sent a reporter over to spend three days with the band, interviewing me, Ryan, and my mom, who had decided to come along.

If anybody doubted my claims about being unfairly treated and the way I was fired from Guns N' Roses, they can now consider the fact that since Axl axed me, he's found an excuse for kicking *everybody* out of the band. He put together a new Guns N' Roses, complete with a cast of unknowns.

I told the *Rolling Stone* reporter, "I'm giving the fans just what Axl offers: one original member and the music they love so much." Whereas Axl's band sounds disjointed and soulless, our band conveys the hungry, underdog spirit that the original Guns N' Roses possessed.

## ⚔ TYPICAL *RS* ⚔

When the *RS* story came out, I was deeply upset. It was basically a "Where are they now?" type of article. I was given a few paragraphs, but unfortunately, they only used the stuff I had revealed about my still being a drug user. We had covered so much positive, uplifting stuff, but for the sake of selling their rag, they focused on the negative. *RS* ended the article with a quote from my mother revealing her constant fear of my suicidal tendencies and what it might lead to: "I dread the day that I receive the *call*."

I was so pissed about the sensationalistic content of the article that I swore off ever speaking to them again. Just a few months later, however, *Rolling Stone* redeemed themselves in my eyes. In their "100 Greatest Albums of Rock 'n' Roll" issue, *Appetite* landed the number 64 spot. They specifically cited my drum technique having a powerful effect on the overall sound of our band.

*You won't get far*
*If you keep on sticking your hand in the medicine jar . . .*

–"MEDICINE JAR," WINGS
(cowritten by Jimmy McCulloch, who died in 1979
at age twenty-six from a heroin overdose)

## ⤙ THE FECAL HITS THE FAN ⤚

In late February 2007, my brother Jamie went to an AA meeting in Beverly Hills. His group has some major celebs in it who share what must be some wild tales of chasing the dragon. But I happen to know that these guys take their sobriety very seriously. It was Jamie's first meeting in a long time. Jamie had been dealing with his own demons, off and on, for about a decade. But that's his business and his own story to tell if he so chooses.

I've always felt pretty shitty about being the guy who turned Jamie on for the first time. Mom went completely ballistic on me when she came over to my house that fateful night to pick him up and saw her pink-eyed muffin all smoked out, sporting a goofy smile. She wanted to kill me, and I couldn't really blame her, although at the time it seemed like no big deal. Every big brother has to deal with a younger sibling's exposure to drugs for the first time. If you ask me, it's better having it happen in your presence, in the safety of your own home, than for it to go down in some dealer's shit hole with a bunch of strangers.

Anyway, by early 2007, Jamie had been in a relapse for nine or ten months. He was greeted with sincere affection mixed with a healthy degree of encouragement to get his ass back on the wagon. Jamie was stunned, completely caught off guard by the presence of Slash. Seeing Slash at an AA meeting is like seeing the Pope at a GNR concert. It just doesn't happen. Slash walked over to Jamie and gave him a big hug. Here was Jamie, GNR's happy mascot, being greeted by one of his gods. Jamie was a one-of-a-kind phenom, the dream-come-true kid who got to hang out backstage with GNR before he was old enough to whack off. Genuine affection was exchanged, and Jamie was very touched by Slash's warm greeting.

According to Jamie, Slash asked how I was doing and Jamie spoke truthfully. He told Slash that I was on my last legs. Jamie reported that I was back to using and abusing with lustful abandon and that after a couple of cardiac episodes and a stroke, my body was in no condition to weather any further assaults.

Now, I'm sure if Slash had asked *me* how I was doing, I would have said that Caro's love had saved me and I was doing better than ever. And I would have believed I was telling him the truth. But the fact was that the boredom of living in Vegas had slowly, subtly, carved away at my soul. I had slipped by degrees over the last three months, and despite Caro's love, I was back to my old routine without realizing it.

Slash told Jamie that they had to do something, because among other reasons, if GNR was ever going to get back together, there was no way Slash would even entertain the thought of a reunion without my ass on the drums. Jamie couldn't believe the band would even consider getting back together, and it was enough to motivate him like never before.

The combination of my brother's love for me and his newfound sobriety helped him get up the nerve to do what he knew was the *only* thing that would ever give me a shot at getting clean. Jamie called it "thug love," which is like tough love to the tenth power. This was the only way they were going to get through to a pig-headed junkie asshole like me.

# 22

# THUG LOVE

Like any good cop, Jamie knew when he needed backup. So on March 19, Jamie, Slash, and a near legendary interventionist, Steve Levy, all met at Burbank Airport and took the two twenty afternoon flight to Vegas. He even had a limo waiting to pick them up when they landed. They all piled in and motored over to the Las Vegas Country Club, where I was now permanently hiding out in my Fortress of Solitude.

Mom had driven over to meet them all in my front yard because she had to see it with her own eyes. She was proud of Jamie and she greeted Slash like a son. She has loved Slash for over twenty-five years and wanted to thank him personally for making the effort.

That's my mom. She is always the first to say thanks and the last to get thanks for everything she's done. But as I write these words, I'm so mad at her that I get all knotted up just thinking about it. Right now I *hate* the fucking bitch. I am furious with the way she's treated me, but that's another chapter—maybe a whole other book.

So after all the lovey-dovey, Jamie told Mom to clear out because he didn't want her near the place when they broke out the heavy artillery and the ugliest of intervention shit started flying. Mom understood this and went back to her condo a few blocks away. Jamie used Mom's key to get into my place and told Slash and Levy to chill in the living room while he went upstairs to fetch me.

Jamie headed straight up to the bedroom because he knew that was where I was kicking it 99 percent of the time. The drug den: my permanent place of worship. It was so funny because Jamie came in and immediately started choking because the air was thick with the smoke I'd been laying down nonstop for like the last fifteen hours: thick rancid smoke from cigarettes, rug burns, bongs, joints, crack pipes, and more cigarettes. And there wasn't one open window for ventilation or sunlight, just the digital glow from my flat-screen and the dozen or so lighters I had on the bed and nightstand.

Jamie was shocked when he first caught sight of me because, surprise, I already looked like a corpse, and you can't save a corpse. I had lost another twenty pounds or so and was down to about a hundred fifteen pounds, which meant I was now lighter than half the chicks I ever fucked. Now, I was just fucking myself.

Jamie threw up his best poker face, then smiled and told me that he was in town to take Mom out to dinner, something I knew he did every month or so. He said he wanted to swing by my place first to say hi and in fact had some friends downstairs with a bag of kick-ass weed. I was out of my bed like a shot, rubbing my palms together in anticipation as we bolted down the steps together.

I flopped on the couch and flashed my best rock star smile. Tasty weeeed time. I said hello to the other two guys and then did a double take. Holy shit. *Slash!* What the fuck was he doing here? Any lingering suspicions or paranoia I may have felt at the time were wiped out by the sight of my brother from another mother sitting there casually smiling at me.

Slash leaned across and gave me a big long hug. When we plopped back down I caught a quick glance of his face all scrunched up like a flummoxed Kermit. I felt a rush of blood to my face. A quick pit check confirmed my worst fears: I hadn't washed in days and must have smelled like the worst combination of stale smoke and rank ass. I felt horrible about that.

As my head cleared a bit from the rush of seeing Slash, my humiliation was quickly supplanted by a growing rage. Wait a minute . . . I realized what my fuck-ass bro was pulling off, or trying to pull off, and the resentment began to build. But I was determined to keep a step ahead of these bastards who had invaded my sanctuary. Before Jamie could begin to conduct the meeting, I started in with some conduct of my own. Bad conduct.

I had some things to say, and I knew that in order to do it, I would have to keep my anger in check, at least until I had vented. So I talked about everything that Slash had done to abandon me and how he never questioned Axl or stood up to Axl for a moment. He never defended me, Steven, the guy who had given him his first guitar. Slash had crashed at my house and eaten my food and basked in my family's unconditional love, and how did he thank me? He thanked me by sticking it to me again and again.

So for the first half hour or so, Slash and the boys just nodded and listened. My voice started to get kind of shrill at the end, and I have to give them a lot of credit for just sitting there and taking it. I'm not sure I would have. I think I'd have grabbed the nearest ice pick and gone to work.

Then it was time for the boys to fire back, and they remained pretty damn emotionless. I have to hand it to them; they were really focused on their little Rambo mission. They wasted no time with their reason for being there. They wanted me to check myself into Eric Clapton's rehab center in the Caribbean. They had cashed in a lot of favors to get me in there. Jamie said he'd help me pack, because we had to be on a plane that night. Tickets had been bought. Plans had been made. Commitments were to be met.

Steve Levy started to say something that sounded very relevant and very interesting, but I couldn't sit there another minute. I raised my hand like a schoolkid requesting he be excused. My nerves were snapping, and I felt light-headed. I know I should have felt the love, but as I got up to go to the bathroom I booted all over the place for like fifteen minutes. It was too much; I think I was in shock. It was like I had to hug Slash to be sure he was real. But even that didn't give me any kind of lasting joy. I just wanted to slip upstairs to my bedroom, get under the sheets, and wait for everyone to just leave.

Particularly Slash; I wished he would please just go. This was the first time in over fifteen years that Slash was in my home, and I couldn't wait for him to leave. The drugs have just screwed me up for good. I don't react to situations the way any sane person should.

I came out of the bathroom and poured out my soul. I told everyone how grateful I was and how this all meant so much to me—that I definitely wanted to check myself into rehab. They hadn't come a moment too soon to save my sorry ass. Thank you. Thank you!

Everyone eyed me like, "Okay, but we know you, Adler. What's the fucking catch?" The truth was that there wasn't any catch. I think that at that moment, I honestly wanted to go. Or at least some part of me wanted to go. But after waiting around for me to finish packing for over three hours, Slash and Levy said they had to get back to L.A. They had families, and unlike me, they had lives. They had dropped everything to come out to my house and show me the love, but it was time for them to head home. They grabbed a cab and took off.

The sad fact was, after the initial rush of seeing Slash, all I could think about was my drug connection coming by my house soon. Like always, I just needed to get high *one more time* before going to the airport. I had successfully dragged my feet long enough to make sure there was no way we were going to make the flight that night. But my goddamn brother must have figured out why I was stalling and that some dealer would soon be making a house call.

Now, there was no way Jamie was going to solo with me overnight and get me on that flight the next day. I was way too slippery for that and he knew it. So he jumped on the phone and started pulling strings. Next thing I know a security guard is rolling up my driveway. This guy gets out of the car, and he's bigger than Texas. I figured he'd have Jamie's back for the duration, and the two of them would hold down the fort overnight and thwart any attempt by me to sneak my drug runners past the gate.

Plus, Texas was packing heat. This was one intimidating prick, but I could see that at the core of things, he was a big teddy bear inside. You just couldn't risk getting on his bad side. Jamie introduced him as Troy and told him I was fucked up, filthy dirty, with festering abscesses all over my body and numerous infections of varying severity.

Jamie said this right in front of me and I could see he didn't care, because at this point, he was becoming livid. Jamie knew I had lost all interest in everything but my next high. So there was to be no more patience, no more understanding; there was only "the Mission," and it was going down with or without my cooperation.

### ⊰ BACKFIRE ⊱

How did Jamie know about the abscesses covering my stomach? Earlier, when Jamie complained that I was stalling, I told him that painful sores on my gut were making me move slowly. I figured it was a convenient excuse I could use to take longer to pack and miss the flight Jamie had booked to Clapton's rehab. Unfortunately Jamie called Mom about the abscesses and found out I also had a blood infection that had recently threatened to travel to my left eye socket. This had been diagnosed a few weeks earlier and luckily the doctors were able to kill the infection. My eye would be fine but I was supposed to have stayed in the hospital for another week so the doctors could thoroughly clean out my blood.

Fuck that. I bolted after three days. One of the nurses came in while I was putting on my shoes. She was completely stunned but managed to ask me what I thought I was doing. I told her I was a hopeless drug addict who had to go home to get loaded, but then I'd be right back. I finished lacing up my shoes and shot out the door. Of course I never went back to the hospital. My mom told Jamie the doctors hadn't had enough time to finish the treatment and that it was possible I could relapse.

After Jamie introduced me to Troy, I did what I always do when confronted with someone who can get between me and my drugs; I turned on the charm. Within minutes, Troy and I were hitting it off like old war buddies. My plan was to get Troy to drop his guard, have a few beers, and watch some TV while I snuck out to the driveway to meet with my delivery boy. But Troy was no fool; he wouldn't let me out of his sight, and the pain in my gut from the abscesses was really starting to act up.

As I came off my high, the pain level went from ten to twenty and I started complaining. I didn't want Jamie to freak any more than he already was, so I told him the pain was just from me tripping into my closet door while packing. Jamie gave me a look like "Oh, puh-lease," and I knew that my number was up.

Troy lifted up my shirt before I could protest and confirmed everyone's worst suspicions. The abscesses had worsened, considerably. There was one open abscess on my stomach the size of a ripe plum, and it needed tending to. Pronto. Too many dirty needles had been stuck into my belly, and now it was payback time.

After seeing this and speaking with Mom, Jamie concocted a plan he knew I'd be helpless to resist. He told me about a friend of his and the house party he was presently throwing. This friend had a phat pad just minutes away, a steamer trunk full of painkillers, and the best weed this side of Negril.

## ⊰ HOUSE CALLS ⊱

I groaned as we headed over to the house party. The pain was coming in waves, hard and fast. I'm no stranger to messing myself up, but this time the discomfort level was spiking off the charts. How do I get into these situations? Why is there always someone around who loves me more than I hate me? Why do they give a shit? Why can't they just let me croak? And when did I start sounding like a whiny little bitch?

Troy and Jamie half carried me into the party house, the main part of an amazing compound ringed with smaller guesthouses, a pool, and a sizable fence. This dude had the sickest house ever, video games, sharks in a massive tank, personal chef, sound studio, pool table, 120-inch flat-screen, the works. Plus Jamie wasn't kidding; this guy had the best weed ever.

I was practically drooling on myself as I was introduced to the play kingdom. The host took me over to the shark tank and I watched him feed the baby sharks. I was loving every minute of it. Then Jamie had the host check out my wretched gut, and before I could clamp my hands over my stomach, this guy took a peek. The situation was serious, but no worries, my host knew an Asian MD in Vegas who made house calls.

Soon, I was completely caught up in video games, and the weed had helped dull the ache some. At midnight, Dr. Feelgood showed up toting two briefcases. He took one look at me and asked me to lay down faceup on the sofa. He removed my shirt and began poking around. He told me I had several advanced-stage abscesses that needed immediate lancing and irrigation. Before I could respond, he asked Jamie and company to hold me down. That's when I got nervous and told him I was fine. Could we please let this wait until tomorrow? Jamie couldn't help but laugh, and the only thing more ridiculous than my request was the fact that I had let the sores get so bad in the first place.

The doctor numbed up my stomach and then started slicing and dicing. The incisions released the poisons, which shot out like little geysers. I started squirming like a baby and Jamie told me to chill; this was the sickest shit he'd ever seen.

This was definitely a new high for lows. He said this beat the infection I had on my arm that had also needed irrigation. The crap that shot out smelled awful, and I noticed that everyone but Jamie and the doc had left the living room. The doc must have shot me up pretty good with painkiller, because I no longer needed anyone to hold me down.

He told Jamie to keep me quiet and still and gave me something to help me sleep. I didn't feel grateful, or relieved, or lucky. My last thought was a wish: I wanted them all to disappear so I could call my dealer and get fucked up.

My condition forced Jamie to put the Clapton rehab strategy on indefinite hold. It took me weeks to get better, and I did nothing to help the situation. Plus I got more and more frustrated. I just wanted to keep partying, but the dividends were becoming smaller and smaller.

While healing up I felt like all the fun had gone out of using. Maybe I was just doing junk to avoid the torture of getting off the shit. I feared that worse than anything. It's such a bitch when your body starts screaming that it wants more *now*! You pump more drugs into yourself, but the high is barely there. They call it "chasing the dragon"; they ought to just call it "chasing the drag."

I got so depressed and fed up with the hopeless hole I'd dug for myself. This must have been one of the all-time lows for me, because I ended up slashing my arms pretty badly. I don't remember doing it, I just remember Jamie's being there to bandage me up and call the hospital. It was horrifying when I tried to recall my state of mind before I did something that might have ended my existence. Even though the wounds weren't mortal, I wondered if I had chickened out or just fucked up.

Now, I knew Jamie was bracing for my refusing medical attention, so I just cashed out by telling him I would willingly check myself in if he'd buy me a dozen Krispy Kreme donuts. Jamie was only too happy to oblige, although he knew I would change my tune as soon as I walked into the ER. He humored me anyway, going along with my Krispy Kreme strategy.

As soon as I had inhaled those shiny sugar bombs, I was having second thoughts. But by then it was too late. He had already told the doctors that I had attempted to take my life (when was I not doing that?). Their policy mandated that they keep me for seventy-two hours of round-the-clock observation.

So there I was, furious but stuck. I planned to sneak out, but they put me in restraints. For three days I went through the most hellish withdrawal, squirming and sweating, my body wracked by a nonstop assault of the worst cramps and chills, the most heinous nausea, and the overall sense that I *was* going to die.

Fuck that. I wanted to die. *Please let me die.*

Jamie showed up on the morning of the day they had to release me. He smiled, having told me he would have my "reward" for going through hell. He handed me a milk shake. What balls. But my bro knows me too well and even though I wanted to whip it at his face, I needed to chug it down even more. It turned out to be the tastiest shake I'd ever drunk, a frosty, thick vanilla frappé from heaven. At that point my body must have been craving anything sweet, because it actually helped settle me down.

Before I knew it, I was waking up in a car, and it was nighttime. "Where the hell am I?" Troy was driving, and he just said that he was taking me home and I should go back to sleep. We would be pulling up to the house pretty soon. For some reason his suggestion to go back to sleep seemed like the most perfectly natural thing to do, so I closed my eyes, and I was instantly out.

The next time I woke up, we were still in the car. What the fuck? I was immediately suspicious, but by then it was too late. Troy and Jamie had tricked me, and we were in North Hollywood. That frappé must have been laced with enough tranquilizer to stop a charging rhino. Fuckers knew that the only way they could keep an eye on me was to enlist a squadron of Jamie's friends to help, and they all lived in L.A., not Vegas.

As we rolled up to the house, I knew I was in for some total hell time, because trying to kick what I was on without medical supervision is not only the most painful way to deal with withdrawal, but it is a guaranteed recipe for failure. It can be very dangerous too, because if it's done too abruptly, it can bring on severe shock. I couldn't blame Jamie and Troy though, because deep down, I knew that I hadn't given them any other choice.

Later, I wondered whether the way they had gotten me to L.A. could be viewed as a federal offense. I have no evidence, however, no way of proving that they had deceived me, kidnapped me, and taken me across state lines against my will. Besides, they were trying to help me, and what's done is done.

## ⊰ HELL HOUSE II ⊱

The first Hell House, at Santa Monica Boulevard and Poinsettia, was the place GNR partied. Now, I was at a totally different kind of Hell House. They were actually trying to wean me off of crack by cutting down my dosages, going from a $100 to $50 to $20 worth of rock a day. I spent the next month or so high (but never high enough) on crack.

Then, when I was near weeping, a complete mess primed for a total breakdown, they would slip me another awesome milk shake that was a spiked concoction that kept me in a haze while I was being shuttled from houses in Hollywood to apartments in Van Nuys. About the only good thing that happened during this time was that the abscesses on my stomach healed up nicely. Other than that, I was the most miserable human on earth.

## ⊰ UGLIEST SHOWDOWN ⊱

Finally I remember Jamie, Slash, and a few other friends who had been a part of this endless, grueling 24/7 ordeal sat me down and asked me how I thought I was doing. They patted me on the back and hugged me. They said they cared about me. Each had earned the right to be in that room, because each had made personal sacrifices to help me. They had all contributed significantly to the latest epic chapter of "Save the Asshole."

I smiled my rock star smile, broad, gleaming, and self-assured. I took my time to look at all their faces, drinking each one in. As I nodded to each hopeful shiny face, I saw each smile back. I saw the love in their eyes. I even put my hand on Slash's shoulder and gave it a squeeze. I had their attention and I had their hopes. And I owed them. God knows I owed them.

But the only thing I owed them was the truth. So I told them that it didn't matter if they kept me there for a month, six months, a year, or a decade. It didn't matter if they tried to wean me off crack, or keep me sedated, or lock me in a room, or chain me in the cellar.

Why didn't it matter? Because as soon as they let down their guard, I was out of there. And I promised them that I would go straight out and do more drugs than I had ever done before in my life. I would be the highest, most fucked-up ingrate in the history of mankind.

That's how I'm doing, fuckers.

Then I saw the light go out of their eyes. I saw their smiles die. I saw some of the dearest people in my life shift in their seats and look down at their hands. I watched those hands clench into fists. I watched their brows furrow and the muscles in their necks tighten. I saw their hopes fade and their resentment flare and I didn't care.

No matter how much anger they felt, it couldn't come near to the raging contempt I felt for myself at that moment. Fuck 'em.

I.

Don't.

Care.

They got up and left. I quickly made a couple of calls, and by midnight, I was back in Vegas, lying on my filthy rug, my heart thumping out of my chest, high on the rock of ages. I really went for it this time, a five-day bender. On the last day I drank so much Jäger I passed out before I could kill myself. And that's why I'm still alive to tell you this story, because I tried to kill myself partying and fucked it up. Again.

I woke up on that rug, lying in a pool of piss, and slowly blinked through the cobwebs. Finally, I focused on a crack pipe not ten inches from my face. It had a healthy amount of rock still nestled in it.

How sweet. Breakfast of Champions.

# 23

# BACK *from the* ABYSS

This went on and on until I heard the crying. Every human being on earth knows this sound, because we all cause it to happen at some point in our lives. It's the sound of our mother crying. And it's the saddest sound in creation. I could hear her as she tossed my clothes into a suitcase and grabbed some things out of my closet and bathroom. I asked her what the hell was going on, and she told me I had called Jamie and uttered one sentence, then hung up. And that was the reason she was in my house. That was the reason she was crying, but they were "happy tears."

I asked Mom what I had said. I didn't remember, but I was definitely curious. My mom isn't surprised by anything I say anymore. She straightened, dabbed her eyes, and said that I had called Jamie to say, "I'm ready now."

To my sheer amazement, I didn't argue or deny what I had said. I felt so fried, so hollowed out, that there was nothing left of me to bitch. I'll be damned, I *was* ready.

I remember looking down at my hands. They hurt so much from lighting and relighting the pipe. I'd get so high I would watch the flame burn down until it was licking my fingertips; but it didn't matter. Butane torches are much more efficient for smoking rock, but when you're all fucked up on crack, it doesn't matter. Nothing matters. You'll light yourself on fire to get high. And that's what I was doing, because my fingertips were all blistered and burned. My mouth felt charred, and I had this horrible chemical taste on my tongue. My head felt cold and empty and my body completely gutted.

I thought, "So this is what it's like." I had *finally* hit bottom. Rock bottom.

I heard my mother pleading with me to hurry up. Her eyesight isn't the greatest and she doesn't like to drive at night. We needed to get on the road if we were going to be in L.A. before nightfall.

Okay, Mom. I just need to hit the bathroom. Incredibly, I was still using. I know that makes no sense, but this is what actually happened. Forty-five minutes later my mom was banging on the bathroom door. I stared at the glass pipe in my hands and figured that with one more good hit, I could make it out to the car and we could get going. I told her I'd be right out.

A half hour after that announcement, Mom was back at the door banging and threatening to call off the whole trip. I guess this got me out of the house, because the next time I was hitting on the pipe I was completely turned around in the front seat of the car. I was leaning over the head rest pretending to look for something in the backseat of the car. But I wasn't fooling Mom for a second.

"Steven, stop that."

"But, Ma, I ain't doing nothing!" It's the same way I talked to her when I was twelve. So there we were, heading down Interstate 15, daylight dwindling along with my once bountiful supply of crack.

t was too cold outside to open the windows in the car, so Mom was begging me to stop "smoking that stuff." I nodded okay, then went right back to hitting it. About twenty minutes later, Mom realized we were completely and utterly lost. She somehow got off the highway, and now we could have been in Bumfuck, Idaho, for all she knew. She was very upset. Poor Mom; I told her not to worry, I'd get her to L.A. I heard her laugh, a shrill unnatural giggle. Uh-oh, maybe Mom picked up a little secondhand smoke!

I told her to take the next right. She looked at me, and I got to hand it to her, because although I was higher than Everest, she somehow knew to roll with it. She followed my directions and within thirty minutes the L.A. skyline was in view. But Mom was fading fast and her eyes were itchy. She wasn't seeing well, she was exhausted, and the smoke was making her miserable.

Mom realized she couldn't check into a hotel that had a lobby. She rightfully believed it would spell doom the first time motel management caught sight of me. We needed to find a place that had an office where you could check in, get your key, then go park your car in the motel parking lot by the entrance to your room. In other words, we needed to locate a roadside fleabag dump right away.

And we did. Soon, I was happily spread out on one of the beds, TV on, reacquainting myself with the pipe. She yelled at me for not helping her unload the car. She was furious. I managed to say I was sorry between tokes. She told me to lower the volume; she'd had it with me. I ignored her.

Things didn't get any better. My mom started the next morning by talking with Jamie, who was doing everything in his power to find a suitable place for me in Hollywood. He and our cousin David were on it, but it's not easy finding a pad that's ideal for a severely impaired drug addict who had to have a pool, a big-screen TV, and a gated entrance.

By the third day, the motel situation had severely deteriorated. My mother was sleeping in the bathtub with the door shut and a wet towel shoved under the door. She was convinced I was going to fall asleep with the TV blaring and a lit cigarette in my hand that would burn down the whole goddamn motel.

By the evening of day three my rock supply was gone. I told Mom I was just going out to get more cigarettes and she had the audacity to follow me outside when I tried to cop. No matter how much I threatened her, no matter how much I screamed, she wouldn't stop tailing me. She stayed right on my ass. She was so obnoxious, I just wanted to punch her the fuck out.

I finally gave up; I couldn't get close enough to any sketchy dealer types without their being scared off by the wicked witch behind me. When we got back to the room, I was so fed up with her that I shoved her, hard. I knocked her down, and there's not a cell in my body that felt any remorse for what I'd just done.

Mom slowly got up. She was shaken but not a bit scared of me. She composed herself and calmly told me she knew it wasn't me who had done that to her, it was the drugs. But it was the last straw for her. She called Jamie, who was at the motel in like ten minutes. He took over the situation and told me my days of pushing Mom around were over.

When I shoved Mom, it was like I was watching a film of this horrible person doing these terrible things. Then at some point in the movie, I caught a reflection of who the person was in a mirror, and it was me. I was surprised, and ashamed, but completely helpless to stop it.

Jamie's spent a good portion of his life telling me what a worthless piece of shit I am. And at some level, he's the one guy who gets through to me. He's the one guy who hurts me when he says those things, not because he's my brother, but because he's my brother and he's right.

At this point, Jamie had pulled out all the stops to find a place for me, exorbitant rent be damned. He quickly checked us out of that motel and by the time we'd packed, cousin David had picked up the keys to a place up in the Hills for about six grand a month. It was not a moment too soon, because Mom was at her wit's end.

## ⚔ ROCK BOTTOM ⚔

They stuck me in a pretty decent crib up near Queen's Road in the Hills, and I was put on a strict round-the-clock watch. The crack pipe got tossed for a bong and an endless supply of Jägermeister. Now, I know to many out there that's anything but clean living, but for me, that's the pure-as–Snow White's–snatch regimen. And the boys knew that with those two substitutions, I might at least have a shot at getting through the first few days without longing for anything stronger.

But before long, the severe devastation that had brought on my desire to call Jamie and announce my readiness to clean up had faded. I wanted to party again. I wanted to get loaded. And I wanted it now.

I began doing everything possible to slip some rock past security. I had my dealers in Vegas ship it to me using a variety of innocent-looking containers. But the bastards in the house intercepted everything. Maybe I should have tried using Snow White's snatch.

## ⚔ *APPETITE'S* TWENTY-YEAR ANNIVERSARY ⚔

The final straw came when I announced to no one in particular that I needed a new microphone. There was a big reunion coming up marking the twentieth anniversary of the release of *Appetite for Destruction*, and I was determined to honor that event with a kick-ass concert at the Key Club. I was getting Adler's Appetite together, and we were going to do some of the choicest songs off that album. It made perfect sense for me to order a special microphone because I would be talking to the crowd, introducing the band, and setting up the songs. No one would suspect that I'd try having drugs stashed in a mike.

When the microphone arrived by FedEx, some fucker intercepted it before I even knew it had been delivered. I kept asking if anyone had seen a FedEx shipment, and no one had. For two days I hounded everyone in the house and cursed out FedEx (who swore it had been delivered but that the confirmation signature was illegible). I swear I was ready to torch the whole fucking house in the hopes of getting a contact high. There had to be some contraband in the deepest recesses of a toilet kit, a jean pocket, or the carpet.

The following morning, I was told the mike had arrived and was in the kitchen. Praise the Almighty. There would be enough crack in there for at least a couple days of partying. My pulse rate shot up, and I began to enjoy the familiar "pre-high" that addicts get right after they score. I walked in, and my jaw dropped. I immediately flew into a rage. There on the table was a carefully disassembled microphone, sitting in about five parts on the table.

No rock in sight. I was beyond furious. If it wasn't for the fact that I had to keep it together for the upcoming Key Club concert, I probably would have done something really desperate.

## ⊰ LIKE A SUICIDE ⊱

I did anyway. I slashed my throat. The guys who were monitoring my every move in the house didn't see that one coming. Well, that's what you get for messing with my deliveries. Fuck 'em. Wait a minute, I'm the one bleeding out!

So on June 13, 2007, I was taken to Cedars-Sinai Hospital bleeding to death. Fortunately I was put into the very capable hands of one Dr. Fine, who ordered I be put under suicide watch. Dr. Fine sewed my neck back together and then sedated me.

Being under their care must have been just what I needed to retrieve my sanity, because I got out of that place with more determination than ever to pull off an epic evening for *Appetite*'s twentieth. I worked with my Adler's Appetite mates to revitalize all the classic songs, and you wouldn't believe who showed up to sit in on rehearsal.

## ⊰ THREE-FIFTHS OF GNR ⊱

Izzy and Duff dropped in on our third day of practice. My heart soared. It felt so great seeing those two shuffle through the door. It was just like old times, the best times (even though I don't remember a lot of them). The boys in Adler's Appetite were only too happy to let the maestros sit in and jam. Now, it was becoming more about who was going to be there rather than who wasn't.

The songs were cooking up fine, and Duff and Izzy sounded great. These were my brothers. My bloods from the days of trench warfare, when no one believed in us but us and my mom, who I must admit is the first and the truest GNR fan ever.

The date for the concert was coming up fast, and it was amazing how quickly we got it together for the big night. It felt great to be back behind the skins. I was still a little shaky when July 28 rolled around, but I wanted to rock so badly, it didn't matter.

## ⊰ TWENTY YEARS! ⊱

The night Adler's Appetite played at the Key Club, we were louder and sounded better than GNR ever did. All right, maybe I'm a little prejudiced here, but we sounded tight. What really made the difference was having Izzy and Duff up there with me. Then Slash walked in. The place, which was already at critical mass, totally erupted. I mean, insanity took over. The only downer about having four-fifths of GNR's original lineup under one roof for the first time in forever was Slash's bewildering decision to *not play*.

Later it got back to me that Slash, in his well-intended wisdom, decided that playing with us would so enrage Axl that it would doom any hope of a future official GNR reunion. Slash had recently gone on record saying that if GNR got back together, it could only be with the original *Appetite for Destruction* lineup, and I think he just didn't want to jeopardize that in any way. If he honestly felt that meant not pissing off Axl, then I have to respect his gut.

And you know what? It didn't matter. It was perfect just the way it came off, feeling the supercharged atmosphere, feeling the unbridled affection of the crowd. I want to take this opportunity to thank Duff, Izzy, and Slash for showing the love that night and getting up onstage with me. Only one thing could top it, and I hope Axl has the room in his heart to make it happen one day.

## ⇥ EURO BABES ⇤

The afterglow from that event actually got me more excited about the Adler's Appetite tour of Europe. I was really up, more psyched than ever when we got to the airport to begin our tour. We were all on the plane, locked and loaded, when someone accused me of being so drunk and disorderly that they had to kick us off. This forced us to miss our connecting flight.

Now, that in and of itself is just standard operating procedure in my life. What really sucked was by the time the booking agent rerouted us all on later flights, we had to pay top fucking dollar, against which our old tickets could not be applied. Someone figured out that we were basically doing the tour for next to nothing. We literally had to go to Europe just to pay for our expenses and break even. I wonder if that's happened before: the Tour to Pay for Our Tour Tour.

But it still felt great to be greeted by the best fans in the world. The enthusiasm and energy, the devotion and love that they put out for every sold-out show was incredible. Nowhere can you find sweeter, happier rockers. They knew every word to every song. They swarmed the stage and screamed for encore after encore. I was getting by on smoking and drinking and wasn't thinking about heavier stuff at the time. I think all that affection kept me from seeking the dark side, at least for the time being.

Whenever I'm onstage, that's the best high, and I realize that's what I'm chasing after the other days of the year. As a rule, GNR audiences are incredible; there's even a great DVD we produced of an earlier tour, this one in Argentina, titled *South America Destruction*.

It captures the insanity of our shows in Rosario, Buenos Aires, and other cities. You can see how thrilled we are backstage, onstage, at every stage of our tour. Take a look at the fans and you'll see the happiest fuckers on earth. The fact that I'm spreading the rock 'n' roll message into my forties makes me loud and proud.

The trouble is that the exhilaration is fleeting, and before the cheers fade away, the meter is already back down to zero. There's no longer any sustain, no longer the thrill of the kill for me. It just feels like I'm staving off the lethal boredom that's always threatening to do me in if I'm not partying or playing music. So it was no surprise that I was quickly back to my self-destructive best by the end of 2007.

## CELEBRITY REHAB

It came as quite a surprise to me, though, when I got an idea that could actually slow my sprint to an early grave. I was watching TV with Jamie in Vegas and I had this moment of clarity. It was that VH1 show *Celebrity Rehab with Dr. Drew*, and Dr. Drew was making some sense, talking to a young addict. I just blurted out that I liked Dr. Drew and that I wouldn't mind being on *Celebrity Rehab* if it meant I could work with him.

Well, you should have seen the expression on Jamie's face. It was as if I had yanked open the thick black drapes in my room and let in the light. He didn't even dare to ask me if I was serious, he just rolled with it, and by the end of the week, David Weintraub and Josh Bender, who work in casting for the show, had brokered a deal to get me on the second season of *Celebrity Rehab*.

Of course they still had to clean me up enough so that I'd be coherent and presentable enough for a steady appearance on the show, and that's when the war of wills started to get ugly. For starters, just try to get me out of my house. My ass was anchored to my home and I wasn't about to go anywhere. And that meant for anything. If they want to have a reality show or a GNR reunion, they can have it in my kitchen, 'cause this dawg ain't leaving the pound.

While Jamie and his pals worked on getting me out L.A. way, I thought this would be a great opportunity to get loaded again, every day, until they needed me on set. I mean, I was willfully going to submit to rehab for who knows how long, and it would have been impolite for me to show up without being in desperate need of help. So I snuck off to call every delivery boy in Vegas to bring me the goodies. But Jamie had headed me off at the pass like never before with a guy who was able to shut me down completely.

We'll call him the Shadow, and he had the instincts, patience, and physical stamina to stay on me like white on rice. For the next couple of months he not only always knew where I was and what I was doing, but he also kept tabs on every delivery boy, gardener, florist, FedEx worker, mailman, dealer, dealer posing as a friend, and anyone else who stepped foot on the property.

I was desperate to get loaded, but the Shadow must have been one of those people blessed with premonitions, because he was there to head these guys off before they could get to the doormat. He was so tenacious, so amazing, that after a while, I couldn't help but give the guy props. And the Shadow kept me off drugs up until it was time for me to show up for the folks at VH1.

## ☧ DR. DREW ☧

Not only do I have a lot of respect for Dr. Drew, but I consider him a friend, someone who is compassionate about helping the addicts he meets and turning their lives around. There are few people in the field whose heads don't fly up their asses as soon as fame knocks on their door. Most celebrities are massively insecure douchebags, but Doc Drew keeps it real. He may be famous, but he hasn't let that go to his head.

Although I wasn't crazy about all the cast members in *Celebrity Rehab,* they all seemed to like me. I was able to open up and really got a lot out of our discussions. But I think the show is basically flawed, because the motivations are false or completely fabricated.

Let's be frank: if you need to be on a television show to try to quit drugs or drinking, or both, your priorities are fucked. If I didn't have a special admiration for Dr. Drew, you would never have been able to get me on the show.

Looking back, I actually made enough progress to not consider the show a complete waste of time. The thing I was most excited about after the show wrapped was my thirty-seven days of complete sobriety. There was also serious talk of a spin-off show, *Sober House,* that would be a reality show follow-up, tracking the progress of the second-season cast members of *Celebrity Rehab.*

## ⇥ ROCKLAHOMA ⇤

Thirty-seven days clean, sober, and excited about playing music again! I couldn't have been more pumped to get the fuck out of L.A. and sit in with some choice bands at Rocklahoma, a music festival being held in Oklahoma. That was a great time for me because I was sharp-minded enough to realize that music could be a key motivator in my life for *staying sober.*

When I'm up there on the stage, playing my heart out and giving it everything I have, I get such an incredible natural buzz. I don't want to disappoint my bandmates with sloppy play, and this begins to extend to everything I'm doing. Stay frosty, maintain your edge, and you will be so much happier. Those few days at Rocklahoma gave me as much or more incentive than all the time I spent on *Celebrity Rehab.* Just have to keep it going, Stevie.

We got word that the *Celebrity Rehab* spin-off, *Sober House,* was green-lit, with yours truly as one of the featured celebs. They were going to shoot it right away, in order to ensure the continuity that they felt was one of the key elements to the look (and success) of the show. Here I was, starring on another national TV show dedicated to my health and well-being. How great is that?

**C**ould things be looking up for Adler? Could a new shot at putting my life back together be in the cards? Could we possibly be more fucking deluded here? No.

Why? Because the first thing that I did when I got back to L.A. was slip back into partying, or as I termed it, "celebrating my being on the new show." And I did that by getting as fucked up as possible. I got so loaded I didn't even notice how Tuesday, the first day of shooting *Sober House,* kind of melted into Wednesday and Thursday. I wasn't sure if I was fooling anyone by that point, but it hardly mattered. That's because *any* chance of my kidding anyone was blown to bits when the producers of the new show decided to throw a celebratory barbecue for the cast and crew.

What gave me away? Maybe it was the way I was slurring my words, being belligerent to the point of punching out walls, and being really abusive to the cast. I was so out of control that they were finally forced to call the cops.

Now, when the police arrived, the way it went down, these guys were extremely cool cops who actually considered letting me off the hook with a stern warning. But then, of course, my legendary shit luck kicked in. Evidently one of the police officers found out that Rodney King was a member of our cast. Well, Rodney's long-running antagonistic relationship with the LAPD totally fucked my goose, because now the cops had to play it strictly by the book. Having the cameras rolling the whole time certainly didn't help the situation. It all meant I would end up in some deep shit, off the show, and in the slammer.

I was put into rehab, but they kept moving me around, first the Las Encinas and then the Pasadena facility. Fortunately, both centers had some relationship with Dr. Drew; he was probably a member of their staff or on their board of directors. My lawyers cut some deal where I did not have to plead to any wrongdoing until August 20, 2008, when I was going to have the rest of my life determined by a judge. Happy days!

On August 20, I went before the judge, who quickly ruled that the best way to deal with me, at least in the short term, was to stick me back in the Pasadena facility. When I exited the courtroom, some reporter asked me what my next project would be.

I answered truthfully. Steven Adler's sobriety is my next project, one day at a time, or as I put it to the journalist, "one cigarette at a time." Staying clean and sober is my present and future project, and that's all I'm working on until I get it right.

Nothing is more important to me now. And I've got a new streak of sobriety going that I'm determined to build upon. I've got to do this for Caro and me. And don't think for a moment that it's only because I've finally seen the light and am trying to change. It's also because there's a nasty ultimatum that's been made very clear to me: one more fuckup, and I could be facing a mandatory three-year jail sentence. That got my attention. Well, maybe.

## ⇥ WRITING THIS BOOK ⇤

Staying clean is all about staying busy, and that certainly was a damn good reason for writing this book in 2009. And if my story can keep one rock 'n' roller away from hard drugs, then *My Appetite for Destruction* will be a resounding success. While I'm sure many rock musicians have given the same lofty reason for writing about their lives, has any kid ever refused to abuse because of what they've read in a book about rock stars? Truthfully? I fucking doubt it. So while that would be nice, my sobriety has taught me to keep it real.

Keeping it real means admitting, at the beginning and end of my story, that I've been a selfish asshole. No apologies. And although I've learned to be less selfish, I realize you've got to please yourself in life. I hate people who go around figuring out how to sacrifice and please others. They usually just end up pissing off the ones they want to please. I say please yourself and you'll please others.

*Appetite for Destruction* is often cited as the immortal soundtrack for a whole generation, but that's mainly because the five of us were only interested in writing songs about our own lives. People thought either it was great or it was shit, but it was *our shit*.

*Comprende?* And I am eternally proud to be the foundation, the pulse, for that amazing soundtrack: my drumming, my beat, my music, my life. No one can take that away.

The last song I ever played on as GNR's drummer was "Civil War." And to borrow a line from that song, I was kicked out because "some men you just can't reach." Well, I want to reach people with this book and show them that when you refuse to deal with life, life deals with you, and it's brutal. I've spent twenty-five years learning that lesson, and I can finally say I've earned the right to quote another lyric from "Civil War": "He gets it."

I get it. And that doesn't mean I won't fuck up again, but I will no longer do it from the dark side. Caro's love is my light and salvation, and I am going to try harder than ever to deserve it and prosper from it. Our bond has finally given me a family that will never kick me out, a family I can embrace and be a part of forever.

## ⇥ ADLER'S APPETITE GROWS ⇤

'm also determined to keep Adler's Appetite together. And I'm getting lots of support from Duff, Slash, and Izzy. As I ease into the next decade, I get all choked up just thinking about jamming with my brothers again. When we got together at the Key Club for *Appetite*'s twentieth, it felt like no time had passed and none of the bullshit had ever gone down.

Everything is going to be great. Staying sober will allow me to tour the world again and again, and thank the millions of GNR fans out there for keeping the faith and never giving up on me. As Freddie Mercury sang in Queen's epic song "We Are the Champions": "I've paid my dues, time after time" and "I thank you all."

It's been no bed of roses—it's been fudge-packed with myopic judges, greedy producers, parasitic managers, scumbag lawyers, fair-weather friends, and shady promoters.

But you know what? It's going to take a lot more than that to ruin my appetite!

# ACKNOWLEDGEMENTS

For your encouragement, inspiration, guidance,
friendship, patience, and love:
Axl,
Duff,
Izzy,
Freddie Mercury,
Roger Taylor,
Adler's Appetite,
Steven Tyler,
Tommy Lee,
Nikki Sixx,
Kevin DuBrow,
Dallas Taylor,
Sammy Alianzo,
David Mancini,
Fred Coury,
Lisa Ferguson,
Sheree Barnes,
Ty Estrada,
Bob Timmons,
Vicki Hamilton,
Leo Garcia,
Mauro DiPreta,
Jennifer Schulkind,
Alan Brinkley,
Alan Heimert,

Dennis Dasher,
Richard Check,
Michael C. Rockefeller,
Katherine V. Spagnola,
Sarafino J. Spagnola,
John and Kathy Spagnola,
the DeNadais,
the diBonaventuras,
the St. Johns,
the Camerons/Richardsons,
the Fleischmans,
the Wiedemanns,
the Swifts,
the Kennedys,
coaches Maio, Brannon, Mazza, and Restic,
Adam Chromy,
the Pinkels,
John and Marion Viglione,
Lorenzo and Constance Spagnola,
Nicole, Megan, and Kelly Spagnola,
David Sears,
Leo Barth,
Freddy Anderson,
and
Steve Botein.